CHEMICAL MUTAGENS
Principles and Methods for Their Detection
Volume 3

Sponsored by the Environmental Mutagen Society

CHEMICAL MUTAGENS
Principles and Methods for Their Detection
Volume 3

Edited by Alexander Hollaender

Biology Division
Oak Ridge National Laboratory
Oak Ridge, Tennessee
and
University of Tennessee, Knoxville

with the cooperation of
ERNST FREESE, KURT HIRSCHHORN,
AND MARVIN LEGATOR

℗ PLENUM PRESS • NEW YORK-LONDON • 1973

Library of Congress Catalog Card Number 73-128505
ISBN 0-306-37103-0

© 1973 **Plenum Press,** New York
A Division of Plenum Publishing Corporation
227 West 17th Street, New York, N.Y. 10011

United Kingdom edition published by Plenum Press, London
A Division of Plenum Publishing Company, Ltd.
Davis House (4th Floor), 8 Scrubs Lane, Harlesden, London, NW10 6 SE, England

Printed in the United States of America

*Dedicated to Professor Karl Sax
on the Occasion of His Eightieth Birthday*

Professor Sax has long been recognized as the founder of the
American school of experimental chromosome cytology.
With his students and his wife, Hally Sax, he laid the
foundations for much of our present knowledge concerning
the effects of radiation and chemicals on the induction of
chromosome aberrations. In a more fundamental sense, he
employed these agents as tools to investigate chromosome
structure and function. His interests and influence have not
been limited to purely academic problems but have extended to relevant social questions as well. Before most
scientists and the public were aware of the hazards of
certain drugs and food additives, Professor Sax was studying
their effects on cells, and long before the widespread concern
with overpopulation, Professor Sax was warning us through
his writing and lecturing of the dangers that lay ahead. It
is hoped that Karl and Hally Sax will continue for many
years to explore the intricacies of chromosomes with that
special combination of insight, zest, and grace that is their
trademark.

Preface

The ready acceptance and wide demand for copies of the first two volumes of *Chemical Mutagens: Principles and Methods for Their Detection* have demonstrated the need for wider dissemination of information on this timely and urgent subject. Therefore, it was imperative that a third volume be prepared to include more detailed discussions on techniques of some of the methods that were presented from a theoretical point of view in the first two volumes, and to update this rapidly expanding field with current findings and the new developments that have taken place in the past three years. Also included is a special chapter by Dr. Charlotte Auerbach giving the historical background of the discovery of chemical mutagenesis.

Methods for recognizing mutagenic compounds *in vitro* are a necessary preliminary step toward arriving at satisfactory solutions for recognizing significant mutation rates in man, which must be done before our test-tube methods of detection can be considered reliable. Two chapters in this volume make important contributions to this problem.

Due to the increasing activity in efforts to perfect techniques for detecting chemical mutagens and their effects on man, it is planned to continue this series of volumes as necessary to keep abreast of current findings.

Alexander Hollaender

Biology Division, Oak Ridge
National Laboratory and
The University of Tennessee

Contributors to Volume 3

N. G. Anderson
Molecular Anatomy Program
Oak Ridge National Laboratory
Oak Ridge, Tennessee

C. Auerbach
Department of Genetics
The University
Edinburgh, Scotland

Alexej B. Bořkovec
Insect Chemosterilants Laboratory
Beltsville, Maryland

Donald Clive
Mutagenesis Branch
National Institute of Environmental
  Health Sciences
National Institutes of Health
Research Triangle Park, North Carolina

G. E. Cosgrove
Biology Division
Oak Ridge National Laboratory
Oak Ridge, Tennessee

W. Gary Flamm
Mutagenesis Branch
National Institute of Environmental
  Health Sciences
National Institutes of Health
Research Triangle Park, North Carolina

W. M. Generoso
Biology Division
Oak Ridge National Laboratory
Oak Ridge, Tennessee

Alain Léonard
Laboratory of Genetics
Department of Radiobiology
C.E.N./S.C.K., Belgium

J. V. Neel
The University of Michigan Medical
  School
Ann Arbor, Michigan

Howard B. Newcombe
Biology and Health Physics Division
Atomic Energy of Canada Limited
Chalk River, Ontario, Canada

James B. Patterson
Mutagenesis Branch
National Institute of Environmental
  Health Sciences
National Institutes of Health
Research Triangle Park, North Carolina

James D. Regan
Carcinogenesis Program
Biology Division
Oak Ridge National Laboratory
Oak Ridge, Tennessee

L. A. Schairer
Biology Department
Brookhaven National Laboratory
Upton, New York

R. B. Setlow
Carcinogenesis Program
Biology Division
Oak Ridge National Laboratory
Oak Ridge, Tennessee

A. H. Sparrow
Biology Department
Brookhaven National Laboratory
Upton, New York

T. O. Tiffany
Molecular Anatomy Program
Oak Ridge National Laboratory
Oak Ridge, Tennessee

A. G. Underbrink
Biology Department
Brookhaven National Laboratory
Upton, New York

John S. Wassom
Biology Division
Oak Ridge National Laboratory
Oak Ridge, Tennessee

Friedrich K. Zimmermann
Department of Biology
Brooklyn College of The City University
  of New York
Brooklyn, New York

Contents of Volume 3

Chapter 29
Repair of Chemical Damage to Human DNA 151
by James D. Regan and R. B. Setlow

Chapter 30
**Tradescantia Stamen Hairs: A Radiobiological Test System
Applicable to Chemical Mutagenesis** 171
by A. G. Underbrink, L. A. Schairer, and A. H. Sparrow

Contents of Volume 1

Contents of Volume 2

CHAPTER 24

History of Research
on Chemical Mutagenesis

C. Auerbach

Department of Genetics
The University
Edinburgh, Scotland

I. INTRODUCTION

The interest in mutation is as old as the modern science of genetics. It is well known that the term "mutation" was coined by de Vries for sudden hereditary changes in *Oenothera;* although we now realize that these changes were not mutations *sensu stricto*, the term has been retained. De Vries was also the first to dream of the induction of "directed mutations" that would provide man with "unlimited power over Nature." This dream has been a major spur to the search for chemical mutagens. In 1914, T. H. Morgan reported on unsuccessful attempts to produce mutations in *Drosophila* by treatment with alcohol or ether. Mann, in 1923, likewise failed to obtain mutations from treatment of *Drosophila* with a variety of chemicals, including morphine, quinine, and a number of metal salts. During the 1920s, mutation research was put on a firm basis by H. J. Muller, who developed the concept of "mutation rate" and devised objective, efficient, and quantitative techniques for its measurement. These techniques proved their value first in the discovery and analysis of the mutagenic action of ionizing radiation, but they were soon applied also to tests of chemicals. Muller himself obtained negative results with several chemicals, but Russian workers (Lobashov, Saccharov, Magrzhikovskaya) claimed success for ammonia, iodine, potassium permanga-

1

nate, and copper sulfate; for the last-named substance, confirmatory data were published by Law in the United States. Although in these experiments the differences between the frequencies of sex-linked lethals in control and treated flies were statistically significant, treatment effects were always small and rarely reached or exceeded 1% lethals. Since at that time the striking differences between spontaneous mutation frequencies in successive broods were not known, one cannot exclude the possibility that the effective treatments had acted simply by altering the rate of sperm utilization. It would be interesting to repeat some of these experiments with a rigorous brood pattern technique.

II. WAR GASES

During World War I, pharmacologists were struck by the similarity between X-ray burns and mustard gas burns. Both heal only slowly and once healed have a tendency to break open again. For X-rays, these clinical features were thought to be a consequence of chromosomal lesions leading to mitotic disturbances. At the beginning of World War II, Dr. J. M. Robson, a pharmacologist in Edinburgh, conceived the idea that mustard gas, like X-rays, might create mitotic disturbances by chromosome breakage. This idea received support from experiments on ovariectomized mice in which treatment with a weak solution of mustard gas destroyed the normal mitotic response of the vaginal epithelium to hormonal stimulation. In cooperation with Dr. Robson, I tested whether also in *Drosophila* mustard gas inhibits mitosis and whether this is due to chromosomal effects. Both questions were answered in the affirmative. Inhibition of mitosis and meiosis was shown in a study of germ cell development: stages requiring nuclear divisions were arrested, while spermiogenesis and maturation of the oocyte remained undisturbed. This preferential effect on dividing cells was later (1946) corroborated by Bodenstein, who studied the effect of nitrogen mustard on amphibian embryos. Applied to adult *Drosophila* males, mustard gas produced high frequencies of sex-linked lethals: from 7 to 24% were obtained in the first three tests. The ability of mustard gas to break the chromosomes was made likely by the high frequency of dominant lethals, and certain by the production of translocations, inversions, and large deletions. Koller showed that mustard gas shatters the chromosomes in pollen grains of *Tradescantia*. Simultaneously, and independent of the work going on in Edinburgh, cancer workers in the United States found that certain nitrogen mustards had carcinostatic activities. We attributed this to the chromosome-breaking abilities which these same mustards had shown in our experiments on *Drosophila*. This interpretation could account for the finding that the bone marrow with its dividing cells is especially sensitive to mustards, a fact that had puzzled the pharmacologists.

In order to see whether mustard gas, like X-rays, acts via separate "hits" on the chromosome, a research worker in my group exposed *Drosophila* males to increasing doses of mustard gas and plotted the frequency of translocations against dose as measured by the frequency of sex-linked lethals. He obtained an almost perfect two-hit curve and concluded that mustard gas is, indeed, a "hit" poison. We realized, of course, that measurement of the dose in terms of sex-linked lethals had serious shortcomings but decided that it was still preferable to dose measurement by the chemical parameters of the treatment. A great amount of subsequent work on dose–effect curves for mutations has con-firmed that—with the exception of X-rays applied to synchronous nondividing cells—dose–effect curves for cellular organisms are of very limited value for conclusions on the molecular action of mutagens. Very recently, the amount of chemical mutagen that attaches to the extracted DNA of treated mice and flies has been determined. While this is a beautiful way for estimating the average effects of single molecular lesions, it cannot be used for dose determination in routine experiments; in these, a genetic parameter will remain necessary.

The similarity between the genetic effects of X-rays and mustard gas was at first sight very striking. Mustard gas produced all those effects that were known to occur after X-ray treatment. In addition to sex-linked and autosomal recessive lethals, dominant lethals, and all types of large rearrangements, it produced dominant and recessive visible mutations, small deficiencies, and somatic crossing-over. The deficiencies were detected cytologically by Slizynski and Slizynska. The spectrum of visible mutations gave no evidence of specificity. Because of this re-semblance to X-rays, mustard gas and related compounds with nuclear effects were called "radiomimetic." Of late, the term seems to have dis-appeared from the literature, and for good reasons. The diversity of mutagenic and chromosome-breaking chemicals is so great that there is no chemical reality in a group term comprising all of them. Moreover, it was obvious from the start that most of the observed similarities must be due to the fact that the genetic material has only limited possibilities of response to whatever injury it has suffered.

Actually, differences between the genetic effects of X-rays and mustard gas appeared quite early on in the work. One was the marked strain difference in response to mustard gas. When males from different strains were exposed together to the same treatment, the strains could be ranked consistently by their yields of dominant lethals. In the most sensitive strain, the males became completely sterilized by a treatment that only caused about 10% dominant lethality in the most resistant strain.

When spermatozoa of the most sensitive and most resistant strains were treated inside the seminal receptacles of females of a third strain, these differences disappeared: clearly they resided, not in the spermatozoa or their chromosomes, but in anatomical or physiological features of the males that carried them. Differences among strains and species in response to chemical treatment have remained one of the major obstacles to estimates of genetic hazards from environmental substances. Even in the same animal, different germ cells may differ greatly in their response to a given mutagen. This was known for X-rays and now was found to apply also to mustard gas.

Two major peculiarities of mustard gas became evident in our first experiments. One was the relative shortage of translocations compared with X-ray doses that yielded comparable frequencies of sex-linked lethals ("mutagenically equivalent doses"). The other was the high incidence of "delayed mutations," whose appearance was separated from the time of treatment by one or more cell cycles. The former peculiarity may well be a consequence of the latter, since two breaks have to be open in the same cell in order to produce a rearrangement. If only one of two potential breaks opens before zygote formation, a potential rearrangement will be lost as dominant lethal. In agreement with this hypothesis, dominant lethals were found to be relatively more frequent after mustard gas treatment than after irradiation. Small deficiencies, which presumably do not require two independent breaks, were not less frequent in chemically treated than in irradiated chromosomes. Whether additional factors, e.g., an inhibitory effect of mustard gas on rejoining ability, play a role in the shortage of rearrangements is still an open question. So is the nature of the delayed effect. In the pre-Watson–Crick era, every F_1 mosaic was attributed to the delayed realization of a potential mutation in a daughter gene or to the delayed opening of a potential chromosome break in a daughter chromatid. Nowadays, we realize that mosaics are the expected outcome of one-strand lesions in DNA, and it is the complete mutations that require explanation. However, not all mustard gas–induced mosaics could have been due to this cause. Pedigree analysis showed that many of them were "replicating instabilities," i.e., potentially mutant loci that can replicate as such over many cell generations and, in the course of this, throw off mutations repeatedly. Such instabilities have since been found after treatment with many mutagens; their nature is one of the most intriguing problems of present-day mutation research.

During World War I, a study of the biochemical action of mustard gas had been initiated by Peters and his group in England. In World War II, several teams of biochemists carried out intensive research into the reactions of mustard gas and other vesicants and lachrymators with biologically important macromolecules. It was found that these substances attack most proteins, including many enzymes, and that hexokinases and other enzymes are especially sensitive to them. One of our first experiments with mustard gas had shown that no mutations are produced in untreated male chromosomes that have been introduced, by insemination, into treated eggs. This made it unlikely that the mutagenic action of mustard gas is mediated by enzyme poisoning. More conclusive proof was provided by the failure of lewisite and chloropicrin—both potent SH enzymes—to produce mutations. This, as well as the "hit" kinetics of mustard gas, made it very probable that its action was directly on the genetic material, which at that time was considered a nucleoprotein. Most geneticists attributed the specificity of the gene to its protein moiety and assumed that the genetic effects of mustard gas and related compounds were due to their known effects on proteins. It is true that Herriott in 1948 published evidence for the unusually high sensitivity of virus DNA and transforming DNA to inactivation by mustard gas. This was, in fact, one of the first pointers to the essential role of DNA in genes, but, like the mutagenic action spectrum of UV, it was not decisive evidence because energy transfer from DNA to protein could not be ruled out.

Most of the genetic results with mustard gas and nitrogen mustards were obtained during the war and were subject to a ban on publication. During 1942 and 1943, they were submitted to the Ministry of Supply in four reports by Robson and myself and in one by Koller and collaborators. In a Letter to *Nature* in 1944 concerning the weakly mutagenic action of mustard oil, allyl*iso*thiocyanate, Robson and I mentioned the existence of much more potent chemical mutagens. Their nature was revealed first in 1946 in another Letter to *Nature*. In the same year, Gilman and Philips published in *Science* their first report on the antileukemic action of nitrogen mustards. These first releases were followed in quick succession by papers from various countries and laboratories. These contained partly a backlog of accumulated classified information, partly new evidence on the ubiquity of the genetic effects of mustard gas and related compounds. During the same immediate postwar period, the introduction or microorganisms as objects of genetic research provided a new and powerful tool for mutation studies.

III. OTHER ALKYLATING AGENTS

The first suggestion to test ethyleneimines for mutagenic activity came to us in a letter from Brown University, Providence, Rhode Island, U.S.A., where a Ph.D. student (now Dr. C. E. Schilling of Knoxville, Pennsylvania) had noticed the pharmacological similarities between the effects of these substances and of vesicant war gases. After having read our first summarizing article in *Science* (1947), he sent us two sealed samples, one of ethyleneimine itself, the other of a derivative. At the time, we were busy with other things and the vials were put away and neglected. In fact, we forgot all about them until, in the early 1950s, reports about the chromosome-breaking and mutagenic abilities of ethyleneimines, epoxides, diazoalkanes, methanesulfonates, and similar substances issued from various laboratories, notably the Chester Beatty Institute for Cancer Research in London. It soon appeared that the mutagenic abilities of most of these substances and of some additional ones had already been established independently by the Russian geneticist I. A. Rapoport. Using the *Drosophila* sex-linked lethal test, he had shown the effectiveness of ethylene oxide, ethyleneimine, epichlorhydrin, diazomethane, diethylsulfate, glycidol, and several others. His results were published in Russian journals in 1947 and 1948, shortly before Rapoport was expelled from the Party, withdrawn from the scientific scene, and kept incommunicado to visiting scientists. Fortunately, he is now working again in the field to which he has contributed so much. It should be noted that Rapoport tested alkylating agents at a time when most geneticists thought of protein as the relevant part of the gene and that he tested them because of their known reactions with proteins. He mentioned that they also react with nucleic acid but considered this less important. His astonishing success in detecting mutagens shows that it pays to forge ahead even in situations for which no molecular model is yet available.

Several points of interest emerged from comparative studies with alkylating agents. The first goes back to the early war years. In 1942, a report from Peters' laboratory to the Ministry of Supply stated that mustard gas analogues with two reactive chlorine groups were considerably more toxic than analogues with only one such group. Subsequent work has confirmed and extended this observation. Especially in cancer therapy, polyfunctional agents are vastly superior to monofunctional ones. This was attributed at first to the hypothetical formation of cross-links between chromatids and later to the proven formation of cross-links between the two strands of DNA. Subsequent experiments on *Drosophila* have not given support to either notion. In comparisons between homologous mono- and polyfunctional substances at mutagenically equivalent doses, the former were not less efficient than the latter, although it took a longer time for their full effect to become expressed. In the production of point mutations in microorganisms, monofunctional compounds may even be superior to their polyfunctional analogues. The second point of interest concerns the sensitivity pattern of the testis to the action of alkylating agents. Brood pattern analysis in *Drosophila,* carried out mainly by Fahmy and Fahmy, and litter pattern analysis in mice have shown that different agents may elicit very different response patterns in the testis of *Drosophila*, whereas the same agent may elicit different response patterns in mouse and *Drosophila* testes. These results add brood or litter analysis to the requirements of environmental mutagen testing. The third finding was of great theoretical as well as practical interest. Clear cases of mutagen specificity were discovered in experiments on microorganisms in which reversion frequencies at various loci were compared after treatment with UV, X-rays, and a variety of alkylating chemicals. These results were truly exciting; they seemed to vindicate de Vries' dream of directing mutation by specific reactions between chemicals and genes. It must be realized, however, that the observed specificity was not of the kind envisaged by de Vries: it consisted, not of the creation of new "forward" mutations, but of differences between the frequencies with which previously induced genic lesions were reversed. This happened in 1953, the incubation period for the acceptance of the Watson–Crick model of the gene. With its general acceptance, hopes for the de Vries type of mutagen specificity became vanishingly small, while specificity of reverse mutations could now be attributed to the ability of a given chemical to revert either frameshifts or specific types of base change. Actually, experiments on *Escherichia coli* in Demerec's laboratory had already shown that this could not be the whole explanation, for the specific response of a given auxotrophic mutation could in some cases be modified by a change in residual genotype. More recent experiments have confirmed and extended these findings. It now is clear that specificity of reverse mutation may be due *either* to specific reactions between gene and mutagen *or* to specific treatment effects on those cellular pathways that are intercalated between the production of a lesion in DNA and its manifestation as a mutant clone of cells. Meanwhile, undoubted cases of specificity for forward mutations have also been reported. Probably, these are always due to conscious or unconscious manipulation of the physiological and biochemical conditions in the treated cells.

IV. NITROSO COMPOUNDS

Although these are late newcomers to mutation studies, they will be mentioned here because of their possible action via alkylation. Rapoport had included nitrosomethylurethane among his tested substances because of its conversion into diazomethane at high pH. He did, in fact, obtain positive results in the sex-linked lethal test with *Drosophila*, but, like most of his data published on the eve of the Lysenko takeover, they remained largely unnoticed until the early 1960s, when Rapoport himself and, independently, various investigators in the West took up the study of nitroso compounds and found several of them highly mutagenic. Whether this is, in fact, due to the formation of diazoalkanes is still doubtful; on the whole, the evidence seems to speak against it. In bacteria, nitrosoguanidine has a strong preference for the production of mutations close to the replication fork of DNA. The possibility that this preference might extend to strands separated for transcription rather than for replication and that it might apply also to other alkylating agents has led to the first successful attempt in directing a mutagen to a particular locus: the lactose operon in *E. coli*.

V. URETHANE   ⟩

We now return to the early war years when the basis for the detection of a new mutagen was usually a chance observation or a scientific "hunch." In 1933, it had been reported that treatment of *Ustilago* with urethane yielded a shift in the ratio of "postreductional" to "prereductional" meiotic segregation. Ten years later, the German botanist Oehlkers recognized this as evidence for an effect of urethanes (mostly ethylurethane but occasionally also propylurethane) on meiosis in flowering plants. He found no effect on chiasma frequency but obtained very high frequencies of translocations in *Oenothera* and other plants, both monocotyledons and dicotyledons. The first report appeared in a German journal in 1943; because of the war, it reached the English-speaking scientists after a delay of several years. Several features of this work are of special interest for the testing of environmental mutagens. One is the lack of correlation between an effect on chiasma formation and the production of chromosome rearrangements; this was confirmed for a variety of chemicals applied to *Oenothera*. Also in *Drosophila,* where most mutagens produce crossing-over (probably somatic) in the male, there is no correlation between the quantitative effects on crossing-over and on mutation or rearrangement formation, nor are the conditions that favor one or the other kind of effect always the same. This should make us cautious in using recombination, at least the intergenic type, as a measure of mutagenicity. Second, the chromosome-breaking efficiency of urethane was shown to depend strikingly on ionic environment. By itself, urethane was only weakly effective in *Oenothera;* dissolved in $\text{M}/200$ KCl it produced an average of 24% translocations. Third, urethane showed a very marked strain and species specificity of action. The same treatment yielded very different translocation frequencies in two reciprocal *Oenothera* hybrids. While point mutations were produced in *Drosophila* and *Antirrhinum,* there were no genetic effects on *Neurospora* even when Atwood's recessive lethal technique was used for scoring mutations and small deletions in the whole genome. These observations point to an indirect action of urethane; there is evidence that, at least in the inactivation of transforming DNA, the active agents are radicals that are produced via hydroxyurethane. Many chemicals may resemble urethane in acting only under certain conditions and in certain genetic backgrounds; in tests for environmental mutagens, we shall have to keep these possibilities in mind.

VI. ALKALOIDS

Oehlkers also tested some of the well-known alkaloids for chromosome-breaking ability. Morphine, atropine, scopolamine, and narceine induced high frequencies of translocation in *Oenothera*. They were equally effective in the same reciprocal hybrids that showed such marked difference in response to ethylurethane. This does not, of course, exclude the possibility that other strain or species differences in the response to alkaloids may be found. In view of the pharmaceutical use of these compounds, it seems important that they should be tested in mammalian cells. In 1960, a different group of mutagenic alkaloids was discovered by A. M. Clark in Australia. These are the pyrrolizidine alkaloids, e.g., heliotrine, which occur in *Senecio* and a number of other plants and cause liver damage in grazing livestock. They are highly effective mutagens in *Drosophila*, the only organism in which, to my knowledge, they have been tested.

VII. INORGANIC SALTS

The third group of compounds tested by Oehlkers were inorganic salts, which were used either by themselves or in combination with organic compounds. The strongly synergistic action of KCl and ethylurethane has already been mentioned. A similar synergism was found for KCl and fructose. It seems likely that ionic environment plays a special role in the production of structural chromosome changes by chemicals; this should be taken into account when chromosomal damage is used as a criterion for the effectiveness of a suspected mutagen. In the plant *Nigella damascena*, the low frequency of EMS-induced structural changes was increased manifoldly by the presence of divalent copper or zinc ions. The low efficiency with which some potent mutagens, e.g., heliotrine or diethylsulfhate, produce translocations in *Drosophila* might conceivably be due to an unfavorable balance of salts in the cells.

In the tests on *Oenothera*, aluminum and ammonium salts produced fairly high frequencies of translocations when acting by themselves. The effectiveness of ammonium salts recalls the early Russian reports on a weakly mutagenic ability of ammonia vapor for *Drosophila*. When, soon after the war, bacteria had been introduced in Demerec's laboratory as test objects for chemical mutagens, a number of inorganic salts were found to produce mutations. Outstanding among them was manganese chloride. Its effects were shown to be strongly dependent on physiological conditions in the cell. It is therefore not surprising that so far attempts to produce mutations by $MnCl_2$ in fungi have been unsuccessful. Recent experiments on bacteriophage T4 have, on the contrary, yielded positive results. Demerec believed that $MnCl_2$ acts indirectly, by potentiation of some naturally occurring source of mutations. The later finding that $MnCl_2$ concentration plays a crucial role in the action of enzymes engaged in nucleic acid replication makes this idea attractive. In this context, it is of interest to recall that, before World War II, Stubbe had found increased mutation frequencies in the seeds of *Antirrhinum* plants that had been grown under conditions of nitrogen or phosphorus deficiency.

VIII. FORMALDEHYDE, ORGANIC PEROXIDES

The last paper by Rapoport to reach the West before the eclipse of Russian genetics was an especially interesting one. It reported the production of high frequencies of sex-linked lethals in *Drosophila* males that, as larvae, had been reared on medium to which formaldehyde had been added. Rapoport tested formaldehyde for mutagenic activity because of its known reaction with proteins. It is improbable that this idea would have suggested itself to somebody looking for a compound likely to react with native DNA. In fact, the molecular action of formaldehyde is not yet understood, although a hypothesis has been propounded. Rapoport's original finding was followed up by Western geneticists. Their analysis revealed a number of very intriguing features that are of special interest for the testing of environmental mutagens. They can all be subsumed by saying that the effect of the treatment is quite exceptionally dependent on environmental and physiological conditions. Mutations are produced exclusively in early spermatocytes of larvae and only on medium that ensures rapid growth and, in particular, contains sufficient adenosine riboside. Even within each spermatocytic cyst, there are correlations between the loci of independently produced mutations that point to further specificities, perhaps related to the replication cycle of DNA. Formaldehyde finds widespread use in our civilization. In the rearing of pigs, formalin-sterilized skim milk is frequently used; in experiments with *Drosophila*, formalin-treated casein was almost as effective as pure formaldehyde. In view of the very high mutagenic efficiency of formaldehyde–food under the correct conditions, it would seem of great practical importance if these conditions could be defined biochemically.

Formaldehyde provides a good example for another possibility that has to be kept in mind when testing for environmental mutagens: the possibility that a given substance may act in different ways when applied by different means. Sobels found that injection of an aqueous solution of formaldehyde into *Drosophila* males was weakly mutagenic, but in these experiments formaldehyde showed neither the potency nor the specificity of treatment by ingestion. Its probable mode of action was suggested by experiments on *Neurospora* in which the very weak mutagenic action of a formaldehyde solution could be potentiated greatly by admixture of hydrogen peroxide, which by itself is not mutagenic. This suggested that, when given as aqueous solution, formaldehyde combines with metabolically produced hydrogen peroxide to form a mutagenic organic peroxide. The production of mutations in *Drosophila* and *Neurospora* by simple organic peroxides, among them the addition product of formaldehyde and hydrogen peroxide, lent support to this conclusion. It linked up with an earlier observation which, 2 years before, had aroused great interest. In 1947, a Texas group of geneticists (W. S. Stone, O. Wyss, F. Haas, and others) reported the production of bacterial mutations through growth in UV-irradiated medium; the active components of the medium were identified as organic peroxides. At present, this work has gained renewed interest through the search for mutagenic components in radiation-sterilized food.

IX. NITROUS ACID

The discovery of nitrite-induced mutations in tobacco mosaic virus by Mundry and Gierer in 1958; the quick succession of papers in which nitrite was shown to be mutagenic for a variety of microorganisms and for the transforming DNA of pneumococci; the study of the chemical reactions between nitrous acid and the purine and pyrimidine bases of DNA; the ingenious way in which Wittman used these findings for the analysis of the genetic code: all this is recent history and need not be detailed here. Much less well known is the fact that, in 1939 and 1940, Thom and Steinberg in the United States reported the production of heritable variants by nitrite in two species of *Aspergillus*. The authors themselves were doubtful whether these variants were true mutants. The description of their experiments indicates strongly that they were

not. Treatment of spores was ineffective. For the production of the
variants, the cultures were grown for several weeks in liquid medium in
which nitrite was the only source of nitrogen and was present in a concentration that provided only about half the nitrogen required for
optimal growth. Variants were discovered by plating bits of mycelium
or vegetative spores on normal medium; they occurred preferentially in
old cultures. The same types of variant were produced regularly by the
same kind of treatment. Some variants had abnormal morphology or
pigmentation. All of them were "injury" variants, with a paucity or
absence of vegetative spores and perithecia. They could therefore not be
tested for segregation. Thus it seems that the treatment had been an
effective way of producing some kind of specific "Dauermodifications."
In the early 1950s, when the mutagenic action of formaldehyde on
Drosophila was under study, J. B. S. Haldane suggested the use of nitrite
as mutagen because, like formaldehyde, it would attack the amino
groups of the gene proteins. This suggestion was tested on *Drosophila* by
some of my colleagues. The first experiment gave an encouraging positive
result, but this could not be repeated, and the experiments were abandoned. From what we now know about the mutagenic action of nitrite,
we may speculate that the one positive result was due to a fortuitously
low pH in the treatment medium.

X. CALF THYMUS DNA AND OTHER MACROMOLECULES, CARCINOGENS

In 1939, the Russian geneticist Gershenson reported that addition
of large amounts of calf thymus DNA to the food of *Drosophila* larvae
yielded not only phenocopies but also transmissible visible mutations.
Geneticists in Russia and the West argued that, if this treatment were
indeed mutagenic, it should also produce sex-linked lethals. The fact
that this did not happen was taken as evidence against Gershenson's
claim. The war interrupted this work. After the war, Gershenson repeated his claim on the basis of additional data, and geneticists in the
West experimented with his technique. The discrepancy between the
production of visible mutations and the absence of sex-linked lethals in
the early experiments was explained when it was found that the X
chromosome is recalcitrant to the treatment, while the second chromosome responds to it by the production of both visible and lethal mutations. This chromosomal specificity, in turn, is probably a consequence
of regional specificities, for lethals in the second chromosome were found
to be due to small deficiencies that tended to cluster in one particular

region. Subsequently, genetic effects—mainly minutes but also sex-linked lethals and translocations—were observed in the progeny of flies that, by injection or feeding, had been treated with a variety of macromolecules. Here, too, regional specificities were found. The possibility that not all chromosomes may respond equally to a given treatment will have to be kept in mind in *Drosophila* tests, especially in tests of macromolecular agents. These genetic effects of macromolecules seem to be due to some unspecific action on the maintenance of chromosome structure at replication; they have nothing to do with transformation. Minutes were produced not only by *Drosophila* DNA, but also by histones, plasma proteins, and, most efficiently of all, polymethacrylic acid. In the production of sex-linked lethals, bovine plasma albumin was about as effective as deoxyribonuclease. At the Eighth International Congress of Genetics in 1948, Strong and Demerec reported on the production of mutations in mice and *Drosophila* by carcinogens, and since then the problem of the correlation between mutagenicity and carcinogenicity has retained its interest for workers in both fields.

XI. PHENOLS

Shortly after World War II, Hadorn in Switzerland developed an ingenious technique for testing mutagens in *Drosophila*. In order to avoid as much as possible difficulties of penetration and of interaction with the animal's metabolism, he applied the chemicals directly to explanted larval gonads (usually ovaries) which subsequently were reimplanted into untreated host larvae and, in a certain percentage of the cases, became attached to the host's genital ducts. Unfortunately, the technique requires much labor and skill and is therefore not suitable for routine tests; otherwise, it might be developed into a host-mediated assay for *Drosophila*. The chemicals tested by Hadorn were colchicine and phenol. The former was ineffective; the latter was highly mutagenic. In the first experiments, 44 out of 241 treated ovaries yielded from one to many second chromosome lethals, and there were also visible mutations. Clusters of identical lethals were to be expected and did occur. In addition, there was a fair number of ovaries with several unrelated lethals, showing the high efficiency of the treatment. On the other hand, there was a surprisingly high frequency of allelism between lethals from different ovaries, indicating that phenol— like certain macromolecules—does not act randomly on the chromosomes. In contrast to what has been found for macromolecules, phenol-induced lethals always were either nonallelic or strictly allelic, and there was no evidence for overlapping deficiencies. This led Hadorn to conclude that specificity was genic rather than regional, a conclusion that is of the greatest theoretical interest. Unfortunately, the work was abandoned after a few years because it had proved impossible to ensure positive results from experiments ostensibly carried out in the same way and on the same strain. Those experiments that were successful yielded the same high mutation frequencies as the first ones, but quite a few gave completely negative results. These did not, of course, invalidate the conclusion that phenol, under given conditions, can be strongly mutagenic. It only meant that the right conditions had not been found. In mutation tests, clear positive results—unless they have occurred in an isolated "freak" experiment—are always more important than negative results. Chemicals that, like phenol, urethane, and formaldehyde, may give positive as well as negative results are of special interest. It is likely that many environmental mutagens will be found in this group. One of the most important tasks of environmental mutagen testing will be to define the conditions under which such a chemical acts as mutagen. For phenols, the problem has recently become acute through the observation that workers exposed to ambient benzene had an unusually high frequency of chromosome rearrangements in their peripheral blood. In plants, the production of chromosome breaks by phenols, particularly pyrogallol, had been reported as early as 1948 by Levan and his coworkers in Sweden. In these experiments, the chromosome fragments very rarely formed rearrangements, so that they gave rise to cell lethality rather than to transmissible genetic damage.

XII. BASIC DYES

At the Twelfth Cold Spring Harbor Symposium in 1947, Witkin reported that acriflavin produces mutations in *E. coli* and that pyronin appears to do the same. Soon after this, D'Amato tested a series of acridines in plants and found that several of them produced chromosome breaks and—rarely—translocations in *Allium* root tips and chlorophyll mutations in barley. Several years later, A. M. Clark in Australia reported that the addition of pyronin B to the medium of *Drosophila* larvae produced sex-linked lethals in male germ cells. In contrast to formaldehyde–food, pyronin–food was found to act on all germ cell stages in the larval testis. It also produced mutations in female larvae and in spermatogonia of adults that had ingested pyronin. There was, however, marked specificity of a different kind. The weak overall effect of the treatment masked extreme sensitivity of a few males, offset by a seemingly complete lack of response by others. In a test that I carried out by the larval feeding technique, the most sensitive male had one visible, one semilethal, and three independent lethals on 20 tested second chromosomes; in the same experiment, 33 males with together 700 tested second chromosomes had no mutation. Attempts by Clark to determine the causes of this hetero-geneity were unsuccessful. Light or darkness during treatment, high or low temperature, varying degrees of larval competition, varying dye concentrations, injection of azide or potassium cyanide into males—a pretreatment that enhances the mutagenic action of X-rays—were all without effect. From the point of view of environmental mutagenesis, it is disturbing to realize that uncontrolled physiological or environmental conditions may create such extreme differences between individual re-sponses to a mutagenic chemical. In the hands of Crick, Brenner, and their associates and subsequently in those of other molecular geneticists, acridine mutagenesis has become an important tool for studying the genetic code and the nature of mutation, of transcription, and of transla-tion. These findings are too well known to require discussion in this article.

XIII. PURINES

In 1948, Fries and Kihlman in Sweden reported the production of auxotrophic mutations in the ascomycete *Ophiostoma multiannulatum* by prolonged immersion of the conidia in solutions of caffeine (0.2%) or theophylline (0.6%). Either treatment yielded about 1% auxotrophs at survival levels of about 0.2%; control mutation frequencies were very low. Qualitatively, the results were especially interesting because they showed a difference between the mutational spectra induced by these two purines on the one hand and X-irradiation on the other. This was most striking for inositol requirers, which formed less than 5% of radia-tion-induced auxotrophs but more than 40% of purine-induced ones. This is one of the rare cases of mutagen specificity for a whole locus or biochemical pathway. Like similar cases that were subsequently reported from the same laboratory, it may rest on selective effects of the plating medium, amplified by the previous treatment of the plated conidia.

One year after Fries' publication, Demerec and his group reported that caffeine produces mutations in *E. coli*. This claim was confirmed by Novick and Szilard, who designed the chemostat for testing bacteria in continuous culture. They added a number of other purines, notably theophylline, to the list of bacterial mutagens. Pyrimidines, including 5-bromouracil, were ineffective under the conditions of these experiments. The possibility offered by the chemostat to test potential mutagens that are likely to act slowly via disturbances of metabolic processes has been very little utilized after these first pioneer tests. In bacteria as in *Oph-iostoma*, purine mutagenesis showed locus specificity of action: of the two loci tested—those determining resistance to phages T5 and T6—the former responded somewhat less to UV but much more to purines than the latter. It would be interesting to analyze the causes of this specificity. A finding of special interest was the discovery of "antimutagens," i.e., substances that, when present in the chemostat, abolished or reduced the effects of mutagenic agents. The most effective antimutagens were adenosine, guanosine, and inosine; these abolished purine mutagenesis completely, reduced spontaneous mutability considerably, but had no effect on radiation-induced mutagenesis. In this context, it is of interest to remember that in *Drosophila* adenosine is a required comutagen for mutagenesis by formaldehyde–food and that caffeine may act either as co- or antimutagen for bacteria and yeast. A study of the role of purine (and pyrimidine) metabolism in spontaneous and induced mutability would seem a promising and interesting field of research.

In 1949, Kihlman started investigations on the ability of purines to break plant chromosomes (mainly in *Allium* root tips). A comparison among 24 compounds showed that these could be divided into two groups: compounds of the first group acted rapidly and strongly on all cells. compounds of the second group were held back by the lipid perinuclear membrane and could act only when this barrier had disappeared during mitosis. This warns to caution when drawing inferences from negative results obtained on nondividing cells. Much of Kihlman's later work has been concerned with caffeine and 8-ethoxycaffeine. The most interesting point to emerge from it was the recent finding that these substances act quite differently in mammalian cell cultures and in plant cells, most of the difference being due to a difference in temperature during exposure. This is another example—and one of topical interest—for the possibility that the same compound may act in different ways when applied under different conditions.

XIV. BASE ANALOGUES

In 1952, Wacker and his colleagues in Germany found that 5-bromouracil (5-BU) can be incorporated into the nucleic acids of *Streptococcus faecalis*. In 1954, Dunn and Smith in England and Zamenhof in the United States reported that 5-BU can replace thymine quantitatively in the DNA of *E. coli*. With the acceptance of the Watson–Crick model of the gene, these results suggested the possibility that analogues of the purine and pyrimidine bases in DNA might be mutagenic. Litman and Pardee were indeed able to show in 1956 that plaque-type mutations were produced in bacteriophage grown in 5-BU-treated bacteria. In these experiments, and in subsequent ones on bacterial mutations, incorporation of foreign pyrimidines into DNA had to be enforced either by biochemical means, such as addition of sulfanilamide to the treatment medium, or by the use of thymine-deficient strains of bacteria. The absence of such "force feeding" in the experiments by Novick and Szilard probably explains their failure to obtain bacterial mutations with 5-BU. Mutagenic purine analogues, e.g., 2-aminopurine, do not require such precautions. Freese categorized the molecular mechanisms by which base analogues can be mutagenic. His concepts of transitions and transversions, errors of replication and errors of incorporation, have become textbook knowledge. Subsequently, they have been applied successfully to other mutagens.

Yet, from the start there were unexplained findings that could not be fitted into this framework. The first was the discovery, by Benzer and Freese, of "hot spots" for the occurrence of spontaneous and 5-BU-induced mutations in phage T4. This suggested an influence of neighboring bases on the reaction between mutagen and base; recent work from Drake's laboratory has shown the existence of such "position effects." Tessman's finding that 5-BU produces mutations in T4 mainly by errors of incorporation, and in the one-stranded phage S13 mainly by errors of replication, suggests an influence of DNA structure on the production of molecular lesions in DNA. An important role of cellular processes in mutagenesis could be inferred from a paper by Fermi and Stent (1962). They reported that the relative frequencies of different mutations in phage T4 depend on the dose of 5-BU and on the presence or absence of chloramphenicol during treatment, and that 5-BU is mutagenically more effective in singly than in multiply infected cells. The discovery of repair processes; the discovery of the mutagenic action of faulty DNA polymerase; the analysis of cases of mutagen specificity that arise *after* the production of lesions in DNA—all these and a variety of other observation show that the oversimplified picture of chemical mutagenesis as a branch of nucleic acid chemistry has to be replaced by a more sophisticated one in which the reaction between mutagen and DNA, although an indispensable step in mutagenesis, is not in itself sufficient to ensure that a mutation will, in fact, arise. Studies of the effects that physiological and environmental conditions exercise on the quantitative and qualitative results of mutagen treatment are likely to play an increasingly important role in mutation research. Not only will they throw light on the mutational process itself and on its modification by chemical treatments, they promise also to uncover—and have, in fact, already started to uncover— unsuspected concatenations between biochemical pathways in normal cells. For the analysis of the underlying phenomena, microorganisms will be the most suitable organisms. For the testing of environmental mutagens, such studies will have to be extended so that they cover also

the effects of cell type, organ, and organism and of the physiological
stage of the organism on those pathways that connect the entry of a
chemical into the cell with the production of a mutagenic lesion in
DNA, and the lesion in DNA with the expression and perpetuation of
a phenotypic change.

XV. EPILOGUE

Some readers may feel that I have been unfair in the allotment of
space to different periods and different areas of research. Heavy selection
is, of course, required for the condensation of over three decades of research into a brief survey, and this selection is necessarily subjective.
Over and above this unavoidable subjectiveness, I have also quite de-
liberately telescoped my account in such a way that the early period,
say up to 10 years ago, has been expanded, while the later one has been
contracted. I have had several reasons for doing this. The contraction
of the later period can be justified by the consideration that this work
will be known to most readers and that good reviews of all its aspects
are available. It is also much more difficult, if not impossible, to review
such recent developments historically, because it is too early to assess
their positions on the future map of the area of research. I apologize to
the authors of many very interesting and important papers which I have
not mentioned or have mentioned only summarily and anonymously.
The opposite reasons have led me to deal so fully with some of the older
work. Most of it will not be known to readers, and some of it is also
difficult of access because it appeared in German or Russian journals of
the prewar period. The importance of this earlier work can now be
assessed from the vantage point of our present much more complete knowledge of what mutations are. Some observations whose meaning remained
obscure at the time of their discovery can now be fitted into an established
framework of concepts. More important to me seem those observations
that have remained neglected because they *cannot* be fitted into this framework. It is these that should provide a challenge to widening our basis
of research. Finally, for me at least, this digging into the past has borne
the usual fruit of insight into some of the general rules that one has to
keep in mind when working on the present strata of the field. It has driven
home the truths that results which fit with a hypothesis do not necessarily
prove it, that similar results may have different causes under different
circumstances, and that important experimental leads may be neglected
because they do not fit into a current theory. It will have become clear
from this chapter that most of the powerful and potentially dangerous
mutagens have been discovered on the basis of a wrong molecular model

or without any molecular model. Those mutagens that were detected through studies on the chemistry and biochemistry of DNA have been of outstanding value for elucidation of the nature of the gene and of the primary step in mutation; from a practical point of view, most of them are much less important, because they act only in specific systems and under very specific conditions. It seems regrettable to me that few mutation workers nowadays have the sense of adventure that prompted the early workers to penetrate into unknown territories without the guideline of a theoretically acceptable prediction.

XVI. REFERENCES

It is obviously impossible to list the many hundreds of individual papers on which this brief historical survey is based. References will be found in the following review articles:

Auerbach, C. (1949), Chemical mutagenesis, *Biol. Rev. (Cambridge)* 24, 355–391.
Auerbach, C. (1967), The chemical production of mutations, *Science 158,* 1141–1147.
Auerbach, C., and Kilbey, B. J. (1971), Mutation in eukaryotes, *Ann. Rev. Genet.* 5, 163–218.

and in the following book:

Drake, J. W. (1970). "The Molecular Basis of Mutation," Holden-Day, San Francisco.

CHAPTER 25

Observations on Meiotic Chromosomes of the Male Mouse as a Test of the Potential Mutagenicity of Chemicals in Mammals*

Alain Léonard

Laboratory of Genetics
Department of Radiobiology
C.E.N./S.C.K., Belgium

I. INTRODUCTION

Falconer *et al.* [1] and Cattanach [2,3] were the first to detect translocations in the F_1 male offspring of animals given a chemical compound. The positive results obtained by these authors suggested that observation of meiotic chromosomes from treated males *(spermatocyte test on the treated males)* or from their male offspring *(F_1 translocation test)* might be a simple and demonstrative method to routinely test potential mutagenicity in mammals. Such studies were made possible by recent technical advances in methods of making meiotic preparations from mammalian testes which provide a large number of analyzable cells and allow the scoring of many cells per animal and per testis.

This review intends to summarize and to discuss most experimental work carried out with chemicals on male meiotic chromosomes. In

*This work was supported by grants from the Fonds de la Recherche Scientifique Fondamentale Collective.

order to point out important features of the action of chemicals, the results will be compared with those obtained with ionizing radiation, which have been reviewed recently. [4,5] Since this book is intended also to serve as a manual for further studies, the principal techniques of preparation are discussed in detail. A survey of the future possibilities of this method for estimating genetic and cytogenetic hazards of chemical compounds concludes this chapter.

II. METHODS FOR DETECTING CHROMOSOME ABERRATIONS

A. Spermatogenesis in the Mouse

The different stages of the cycle of the seminiferous epithelium in the mouse have been elucidated by Oakberg, [6] using the periodic acid–Schiff technique. Yet although many data on differentiating spermatogonia have been accumulated, the models proposed for spermatogonial stem-cell renewal are still disputed. According to Clermont and Bustos-Obregon, [7] five classes of type-A spermatogonia exist, the A_0 "reserve stem cells," which normally do not divide, and the actively renewing types A_1–A_4. Successive divisions of the A_1 spermatogonia yield intermediate spermatogonia or—in a small percentage of cases (20%)—type-A spermatogonia again, which initiate a new cycle. The A_0 reserve stem cells repopulate the testes after exposure to noxious agents such as ionizing radiation. Oakberg [8,9] and Huckins, [10] however, believe that A_0 cells are normally involved in germ-cell renewal and produce the other type-A spermatogonia. Whatever model is adopted, the most important feature is common to all—namely, that type-A spermatogonia constitute a permanent population, with the transient population formed by the more differentiated germ-cell stages. Therefore, heritable changes induced in spermatogonia by treatment with chemical or physical mutagens appear to be of the greatest interest. Intermediate spermatogonia divide to become type-B spermatogonia, and these again divide into spermatocytes I, which by meiotic divisions given rise to spermatocytes II and spermatids. The latter cells finally differentiate to spermatozoa.

The duration of spermatogenesis in the mouse has been estimated on the basis of experiments with acute radiation exposure or with cells labeled by ^{14}C-adenine and tritiated thymidine. [6,11–13] The following values are generally accepted for the time interval until the cells appear in the ejaculate: spermatozoa in the vas and the epididymis, 1–7 days; testicular sperm, 8–14 days; spermatids, 15–21 days; spermatocytes, at least 22 days; and spermatogonia, at least 36 days. From these values, it can be inferred that by choosing appropriate intervals of time between treatment and mating one can observe in the F_1 offspring mutations or chromosome aberrations produced by treatment of a specific male germ-cell stage.

All 40 chromosomes of the mouse are acrocentric. Their size-range ratio is 5:2 (longest to shortest one), with relatively little difference except length to help in their identification, and even the criterion of size is not very useful since differences between individual pairs are too small. The identification of the sex chromosomes is possible only on the basis of the uptake of tritiated thymidine as determined by autoradiography, since one X chromosome in female cells and the Y chromosome in male cells normally replicate late during S phase. Observation of spermatogonial metaphases (Fig. 1) is therefore suitable only for the detection of numerical changes such as heteroploidy, monosomy, or trisomy, all of which are very rare, or for the diagnosis of certain unstable structural changes such

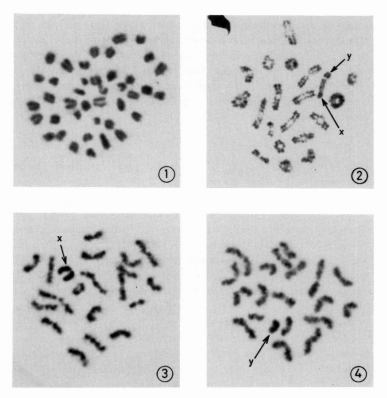

FIGURES 1. Dividing spermatogonia, *2.* Dividing spermatocyte I, *3.* Dividing spermatocyte II with an X chromosome, and *4.* Dividing spermatocyte II with a Y chromosome.

as deletions, which are normally selectively eliminated in the first mitotic divisions after treatment. Such observations appear, however, useless to provide information on the more important stable structural rearrangements. Conditions are more favorable when meiotic cells are considered. During synapsis, beginning at the zygotene stage of meiotic prophase, the chromosomes pair by their homologous segments. These associations give rise to 19 autosomal bivalents, whereas X and Y display a figure resembling an exclamation point (Fig. 2). Autoradiographic investigations with tritiated thymidine have demonstrated[14] that at diakinesis the X and Y chromosomes are associated at the distal ends of their long arms. The first meiotic division yields spermatocytes II, with 19 autosomal univalents and one X or Y chromosome (Figs. 3 and 4).

B. Description of Methods

The characteristic meiotic configurations of spermatocyte I chromosomes at the diakinesis first metaphase stage thus provide a reliable technique for the detection of rearrangements or other chromosome aberrations induced in spermatogonia of the treated males (*spermatocyte test on treated males*). Cytological analysis of dividing spermatocytes can also be used to confirm in F_1 males the presence of translocations indicated by semisterility, the F_1 females being indirectly tested by examination of sons (*F_1 translocation test*) (Table 1).

1. Spermatocyte Test on Treated Males

The following types of structural and numerical changes can be observed in dividing spermatocytes I (Figs. 5–12) after treatment of spermatogonial stages.

a. *Reciprocal Translocation.* Reciprocal translocation involves exchange of terminal segments between nonhomologous chromosomes and requires two breaks and repairs. When the homologous chromosome segments are paired at synapsis, heterozygoty for a reciprocal translocation can give rise either to a ring-of-four configuration (RIV) or, when one or more arms fail to associate, to chain-of-four (CIV), to chain-of-three plus univalent (CIII + I), to two bivalents, to one bivalent plus two univalents, or to four univalents.

More complex aberrations such as hexavalents, octovalents, or decavalents can be observed if two, three, or four translocations occur. Some translocations tend to cause nondisjunction, which results in the presence of an extra element in spermatocytes I. Spontaneous translocations are extremely rare in the control adult males (less than 0.01%) but have been found to increase slightly in old animals. [15]

TABLE 1. Methods for Detecting Chromosome Aberrations

Test	Number of males to be	Number of cells to be analyzed per	Interval of time (days) between treatment and
Spermatocyte test on the treated males	Examined 10	Treated male 100	Observation 50–100
F1 translocation test	Mated 10–20	F1 male 25	Mating

	Mature spermatozoa	Testicular sperm	Spermatids	Spermatocytes	Spermatogonia
	1–7	8–14	15–21	22–35	35

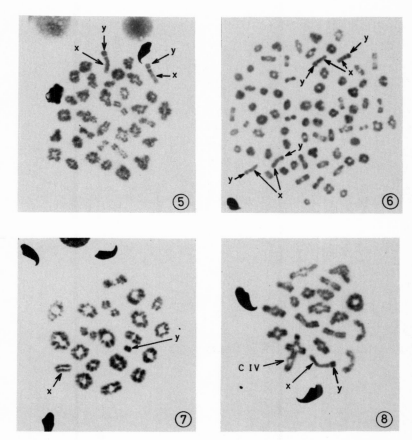

FIGURES 5. Tetraploid dividing spermatocyte I, *6.* Octoploid dividing spermatocyte I, *7.* Dividing spermatocyte I with X-Y univalents, and *8.* Dividing spermatocyte I with 18 bivalents and a chain quadrivalent.

     b. Miscellaneous Structural Changes. Spontaneous centric fusions of acrocentric chromosomes (Robertsonian translocations) have been reported in some strains of mice.[16-18] These structural changes are normally recognizable in dividing spermatocytes I of male homozygotes by the presence of a large ring-shaped bivalent. Such a rearrangement has never been observed in the spermatocytes of control animals having 40 chromosomes. No example[19] of an inversion, whether pericentric or paracentric, or of duplication and deficiency has ever been identified by meiotic chromosome observation in the mouse. However, as shown by the occurrence of quadrivalents in meiosis of male mice carrying the Cattanach's translocation, this technique would be reliable to detect insertions.

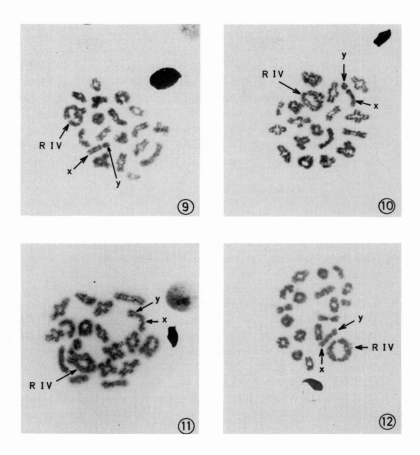

FIGURES 9–12   Dividing spermatocyte I with 18 bivalents and a ring quadrivalent,

    c.   Fragments. Most unstable chromosome aberrations are selectively eliminated during the mitotic process. However, some small acentric fragments stained like chromatin material can be observed in dividing spermatocytes.

    d.   X-Y and Autosomal Univalents. Autosomal univalents resulting[19] from asynapsis (absence of zygotene pairing between homologous segments) or from desynapsis (separation due to the failure to form chiasmata) are rare in control animals, since the chiasmata, formed in the diplotene stage, normally maintain an association of the pairs until the end of metaphase I. X-Y univalents are much more frequent, because the X and Y chromosomes which appear to associate at the distal ends of their long arms do not have homologous segments. In our experiments, the frequency of spermatocytes with X-Y univalents varied from 0 to 10% in most control animals, a value which is in agreement with the 6.17% observed by Lin *et al.*[20]

e. *Polyploidy.* Apparent tetraploid or even hypertetraploid primary spermatocytes are frequently recorded in meiotic preparations from experimental or control mice. Although apparently polyploid spermatogonia are also observed, Ford and Evans[21] conclude from their analysis of T26H translocation that most polyploid spermatocytes are artifacts arising from a fusing together naturally or from a spreading together of adjoining cells during preparation. In control animals, the rate of apparently polyploid spermatocytes varies from 3%[20] to about 6%.[22]

2. F_1 *Translocation Test*

In a translocation heterozygote, alternate and adjacent-1 disjunctions normally occur and give by random segregation at second metaphase 25% balanced original gametes, 25% balanced translocated gametes, and 50% unbalanced gametes. Since adjacent-2 disjunction can also take place,[23] more than 50% unbalanced haploid gametes could be eventually produced. Most translocated–balanced gametes give rise to semisterile offspring. Unbalanced gametes are transmitted as well as balanced ones[23]; they cause death of the zygotes usually around the time of implantation but sometimes even later. Therefore, "heterozygosity for a reciprocal translocation may be suspected from evidence of hereditary reduced fertility."[19] Koller and Auerbach[24] and Koller[25] were the first to detect cytologically the association of four chromosomes at synapsis in three partially sterile lines, and examination of meiotic chromosomes is now extensively used to confirm the presence of translocations in the F_1 sons of treated animals or in the sons of the F_1 females.

3.   *Validity and Interest of the Methods*

If gene mutations are the main genetic hazard against which we have to guard,[26] chromosome damages such as reciprocal translocations which can be transmitted to the progeny through germ cells have to be considered as serious risks, since half of the gametes (the unbalanced ones) from a translocated animal induce dominant lethality in the resulting offspring and half of the survivors still carry the translocation and are normally semisterile. Furthermore, some translocations, probably associated with a position effect, a gene mutation, a minor duplication, or a deletion, can cause sterility of the animal. Certain abnormalities in man are also known to result from viable translocation derived from one parent.[5,27]

Observation of dividing spermatocytes for translocations (spermatocyte test on treated males) induced in spermatogonia, which constitute the main germ-cell type at risk, gives maximum chance of ascertainment before selection has eliminated a proportion of translocations. In that respect, it is preferable to the F_1 translocation test. It is also much cheaper than the specific-loci method, since it gives good results with relatively few animals. Finally, the spermatocyte test is simpler and quicker than the dominant lethal test applied to spermatogonia, and the spontaneous translocations are very rare, whereas the control values in dominant lethal tests exhibit great variation. It must be pointed out, however, that identification or interpretation of some translocation configurations can be somewhat subjective.

The F_1 translocation test appears of particular value for testing female germ cells and postmeiotic male germ cells and can be used to compare the differential sensitivity of various germ-cell stages. For this purpose, the F_1 translocation test is much cheaper and quicker than the specific-loci method. It has the advantage over the dominant lethal test in that it allows demonstration of the nature of the mutational event scored (chromosome breakage) and recovery of the mutations produced for subsequent analysis. Moreover, the F_1 translocation test can verify whether dominant lethals are due to chromosome breakage and permits checking for site specificity of mutagenic damage.

C. Techniques for Preparation of Meiotic Chromosomes

Two basic procedures have been developed for studies on mammalian chromosomes during the last decade: the squash technique and the air-drying method. Both are direct adaptations to testicular material of similar methods used for somatic tissues.

1. Squash Techniques

From 1956[28] to 1964, the squash techniques were widely used for the study of male meiotic chromosomes of small mammals[29–32] and of man.[33] They can yield good preparations which allow counting and identifying chromosome associations. A great disadvantage of these methods is the amount of heterogeneity between samples as a result of the presence of local clones derived from single spermatogonia. Therefore, many slides must be examined to get valid results.

Procedure for Small Mammals[29,32] :

1. A testis is removed, placed in hypotonic sodium citrate solution (0.7% wt/vol) at room temperature, and swirled in the solution to remove adherent fat.
2. The tunica is incised, and the mass of tubules is transferred into

fresh hypotonic solution for 15–20 min or in distilled water for 5–10 min.

3. The tubules are transferred into fixative (50% glacial acetic acid).

4. Small pieces of tubules (1–2 cm length) are placed on a slide and stained with a drop of lactic–acetic–orcein.

5. The preparation is squashed after 10 min.

2. Air-Drying Techniques

Evans *et al.*[34] developed an air-drying technique for meiotic preparations from mammalian testes which avoids certain shortcomings of the squash technique, since it provides a large number of analyzable cells and allows the scoring of many cells per animal and per testis. Since the spermatocyte population from a single testis is thoroughly mixed, the samples of cells scored can be considered random and representative of the whole testis. Furthermore, as pointed out by Evans *et al.*,[34] "the technical quality is, on the whole, very uniform from cell to cell and there is much less cell breakage than with squash techniques." Similar or related techniques have been proposed for somatic tissues[35–37] and for meiotic chromosomes.[38–40]

Procedure for Mouse Meiotic Chromosomes[34,41]

1. Each testis is placed in 2.2% (wt/vol) trisodium citrate solution ($C_6H_5Na_3O_7 \cdot 2H_2O$) in a 6-cm-diameter petri dish. The tunica is pierced and the testis swirled to wash away fat.

2. Each testis is transferred into 2.5 ml of 2.2% citrate solution in a fresh petri dish, and the tunica is removed. The tubules are held with a fine forceps and their content teased out with a curved forceps. Teasing has to be done gently to avoid fragmentation of tubules.

3. The supernatant is transferred into a 3-ml conical tube and centrifuged at 500 rpm for 5 min.

4. The supernatant is discarded, and the pellet is resuspended in 2 ml of 1% citrate solution added drop by drop. The pellet is left in the hypotonic solution for 10 min.

5. The preparation is spun for 5 min at 500 rpm. As much of the supernatant as possible is removed. The cells are resuspended by flicking the tube so that a thin film of cells covers the wall of the tube. Fixative is added (3 parts ethyl alcohol to 1 part glacial acetic acid plus eventually a trace of chloroform) drop by drop, flicking the tube vigorously in order to get a good suspension.

6. After 5 min, the cells are centrifuged and resuspended in fresh

fixative. The change of fixative is repeated after a further 10 min.

7. The final suspension is allowed to stand 1 hr before the preparations are made.

8. With a small pipette, some droplets are allowed to fall on a grease-free slide (stored in a mixture of 1:1 ethanol: ethyl ether).

9. The preparations can be stored or stained immediately with lactic–acetic–orcein, toluidine, or Unna blue. Toluidine blue is generally preferred for immediate examination of non-permanent preparations. Lactic–acetic–orcein and Unna blue are used for finer examination and for permanent preparations.

Careful attention must be given to the quality of the distilled water used for citrate solution. Best results are achieved with water bidistilled on quartz apparatus and with fresh citrate solutions (prepared once a week and stored in the refrigerator). With the technique, one can usually prepare three slides from each testis, each containing from 50–200 cells suitable for scoring or chromosome analysis. An adaptation of this technique involving softening in gluconic acid has been proposed by Eicher, [42] but this method is tedious and the quality of the preparations is generally poor. Therefore, the method of choice appears to be that of Evans *et al.*, [34] at least for mouse and rabbit meiotic chromosomes. This technique has been less succesful in other species, and a simple method described by Meredith [43] gives good results for many stages of male meiosis from most mammals. Unfortunately, this method is rather time consuming, and, as with squash techniques, it is necessary to examine many preparations to avoid sample heterogeneity.

Procedure for Meiotic Chromosomes from Other Mammals [43]

1. Testis tubules are washed in 1% citrate solution and then transferred two or three times into fresh hypotonic solution. The total amount of hypotonic treatment varies from 8 min for rat, Syrian hamster, rabbit, and human testis to 12 min for mouse and guinea pig.

2. The tubules are transferred to fixative (3 parts of ethyl alcohol to 1 part of acetic acid) for at least 15 min; if needed, the tubules can be stored at this step in a stoppered glass tube for several weeks at low temperature (0–4°C).

3. Pieces of the tubules are then transferred to a glass tube containing 0.5 ml of 60% acetic acid and gently flicked until the tubules become transparent as the spermatogenic cells fall into suspension.

4. With a micropipette, one drop is put on a clean slide kept at about 60°C on a warming plate. By drawing in and expelling

the drop several times from the pipette, the drop is "walked" on the slide and then discarded. The procedure is repeated with three or four drops.

This technique provides good meiotic metaphases but takes much time. A similar method has been described by Dyban.[44] A modification of Meredith's method recently reported by Hoo and Bowles[45] provides an increased number (up to 150 suitable spreads per slide) of spermatogonial metaphases and has the further advantage that it results in a random mixing of the individual cells, as is the case in the Evans technique.

3. Conclusions

Different criteria have to be considered to guide the choice of a technique for preparing chromosomes of mammalian male germ cells. First of all, the method to be used will depend on the species to be studied and on the type of chromosome aberrations to be identified. Furthermore, certain methods are time consuming and/or involve the examination of many slides to get a large number of analyzable metaphases and to have a sample representative of the whole testis. In view of these considerations, the following methods are recommended:

a. The air-drying method of Evans et al.,[34] which gives a reasonable random sample of the cell population for the spermatocyte test on treated male mice.

TABLE 2. Translocation Configurations Recorded after Irradiation of Spermato-

Reference	X-ray dose (r)	One translocation			Two translocations					
		RIV	CIV	CIII + I	RVI	CVI	2 RIV	2 CIV	RIV CIV	RIV CIII+I
Léonard and Deknudt[46]	400	66.4	25.7	0.4	2.1	0.2	2.1	0.2	2.5	
Léonard and Deknudt[47]	100–600	73.6	21.8	0.8	1.0	1.0	1.0	0.2	0.4	
Léonard and Deknudt[48]	25–100	74.4	18.3	2.8	1.8	0.9	0.9			
Léonard and Deknudt[49]	250–1250	64.0	24.6	0.6	4.5	4.2	0.9			
Léonard and Deknudt[15]	600	59.7	27.5	0.3	3.9	0.9	2.0	0.9	2.4	
Léonard and Deknudt[50]	250–500	65.9	25.6	1.0	2.7	0.9	1.0	0.2	1.3	< 0.1

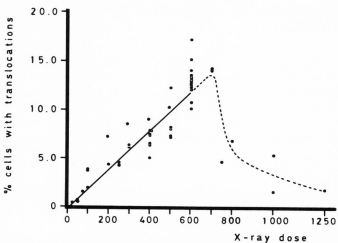

FIGURE 13. Dose–response relationship for translocations induced in spermatogonia by X-irradiation.

b. The air-drying method of Meredith[43] for the spermatocyte test on treated males of other species. This method is also useful in the F_1 translocation test for sterile mice, which may have few germ cells, since it provides more gonial mitoses for inspection than the other methods. This is important when meioses are absent.

gonial Stages (%)

			Three translocations							Four translocations			Five translocation
RVIII	CVIII	RVI RIV	CVI CIV	CVI CIII+I	RVI CIV	RVI CIII+I	RIV CIV	3 RIV	RIV 2 CIV	RX	RVI CVI	RVIII CIV	RVIII CVI
0.2								0.2					
0.2													
0.9													
0.9		0.3											
0.3	0.2	1.0	0.3	<0.1	<0.1	<0.1	<0.1	<0.1	<0.1				
0.3		0.3	<0.1		<0.1		<0.1			<0.1	<0.1	<0.1	<0.1

III. EXPERIMENTAL RESULTS WITH IONIZING RADIATION

Recently,[4,5] the data on chromosomal aberrations induced by ionizing radiation have been reviewed. This discussion can therefore be confined to those results which allow direct comparison with the mutagenic action of chemicals.

A. Spermatocyte Test on Treated Males

After irradiation of the spermatogonial stage, most (about 70%) of the quadrivalent configurations seen in dividing spermatocytes I are in the form of a ring, about 25% are in the form of a chain, and less than 1% are in the form of a trivalent plus univalent. Some cells can be found with two, three, four, or five translocations. No evident correlation is observed between the X-ray dose and the complexity of the chromosome rearrangements[15,46-50] (Table 2). Figure 13 demonstrates that the frequency of spermatocytes with translocation configurations increases linearly from 25 r up to 700 r of X-rays and decreases abruptly after higher doses.[47-49,51-53] Reciprocal translocations require two breaks followed by rejoining; one should thus expect a concave dose–response curve. The expected square law curve can, however, be modified to a pseudolinear one or one with a decrease at high doses as a result of selection by interphase death, early elimination of severe abnormalities, or delayed elimination of small genetic detriments.[54] It should be pointed out that different animals given the same treatment or the two testes of the same animal may display great heterogeneity due possibly to the preferential clonal proliferation after irradiation[55] of spermatogonia showing differential radiosensitivity. In our work with ionizing

TABLE 3. Minimum and Maximum Incidence per Animal of Spermatocytes with Translocation Configurations after Spermatogonial Irradiation

Reference	X-ray dose (r)	Spermatocytes with translocation configurations	
		Minimum (%)	Maximum (%)
Léonard and Deknudt [46]	400	4.5	14.5
Léonard and Deknudt [47]	100–600	1.5	15.0
Léonard and Deknudt [48]	25–100	0	4.0
Léonard and Deknudt [49]	250–1250	0	18.0
Léonard and Deknudt [15]	600	0	24.0
Léonard and Deknudt [50]	250–500	1	12.5

radiation, a maximum of 24% and a minimum of 0% spermatocytes carrying at least one translocation have been observed in irradiated males (Table 3) There is no significant difference among the various strains of laboratory mice with respect to the frequencies of translocations induced by irradiation of spermatogonial stages.[46] It is of interest that the incidence of translocations in the irradiated males decreases when the time interval between treatment and aberration is longer than 200 days.[15]

Types of aberrations other than translocations are much less frequent. Small fragments are sometimes observed in dividing spermatocytes of mice given X-irradiation, but their incidence is always very low.[15,46–48] The mean number of X-Y univalents remains unaltered in most animals, although a few individuals can display an increase to as much as 40%. The causes and the meaning of this high incidence are so far unknown, and no relation seems to exist between the type of treatment and the number of animals with a high rate of X-Y univalents. Finally, the incidence of polyploidy seems not to increase after irradiation.

B. F_1 Translocation Test

Analysis of the F_1 offspring from irradiated spermatogonia [56,57] has shown that dominant lethals and translocated offspring are only about half as frequent as would be expected from the yield of translocations seen in the spermatocytes of the irradiated animals. The frequency of translocation heterozygotes in F_1 offspring after paternal spermatogonial irradiation is therefore always very low (Table 4) but should not be neglected, since, as discussed previously, spermatogonia represent the permanent stem-cell population. When the frequency of translocated offspring in successive matings of irradiated males is used to estimate the

TABLE 4. Genetic Radiosensitivity of Male Germ Cells [58–60]

Cell stage	Mutation		
	Translocation per $r \times 10^5$	Dominant lethality per $r \times 10^4$	X^M0 per $r \times 10^5$
Spermatozoa	25.4	18.0	2.0
Spermatid	72.5	38.0	4.3
Mature sperma-tocyte	7.4	28.0	1.0
Immature sperma-tocyte	20.8	24.0	29.33
Spermatogonia	0–3.6	1.1	0.07

TABLE 5.   Analysis of the Spermatocytes of 43 F₁ Translocated Mice[4]

Transloca-tion	Chromosome associations recorded (%)						
	20II	18II + RIV	18II + CIV	18II + III + I	16II + RIV RIV	16II + RIV CIV	16II + CIV CIV
T2Ald	9	83	8	—	—	—	—
T3Ald	33	60	7	0.001	—	—	—
T4Ald	18.4	50	31	0.6	—	—	—
T5Ald	51	2	47	—	—	—	—
T6Ald	60	33	7	—	—	—	—
T7Ald	1	—	99	—	—	—	—
T8Ald	8	72	20	—	—	—	—
T9Ald	11	67	22	—	—	—	—
T10Ald	4	61	35	—	—	—	—
T11Ald	25	40	35	—	—	—	—
T12Ald	88	—	12	—	—	—	—
T13Ald	34	16	50	—	—	—	—
T14Ald	11	22	65	2	—	—	—
T15Ald	75	6	19	—	—	—	—
T16Ald	18	66	16	—	—	—	—
T17Ald	—	1	79	20	—	—	—
T18Ald	15	18	67	—	—	—	—
T19Ald	—	96	4	—	—	—	—
T20Ald	17	50	33	—	—	—	—
T21Ald	(1)	—	2 + (97)	—	—	—	—
T22Ald	14	19	67	—	—	—	—
T23Ald	19	74	7	—	—	—	—
T24Ald	15	71	14	—	—	—	—
T25Ald	—	19	58	23	—	—	—
T26Ald	6	57	35	2	—	—	—
T27Ald	62	15	23	—	—	—	—
T28Ald	13	—	87	—	—	—	—
T29Ald	8	1	91	—	—	—	—

T30Ald	40	26	34	—	—	—	—
T31Ald	8	1	91	—	—	—	—
T32Ald	12	78	10	—	—	—	—
T33Ald	2	89	9	—	2	21	22
T34Ald	8	18	29	—	—	—	—
T35Ald	7	80	13	—	—	—	—
T36Ald	10	36	54	—	—	—	—
T38Ald	29	66	5	—	—	30	—
T39Ald	2	61	7	—	42	36	—
T40Ald	—	15	3	—	—	—	4
T41Ald	5	83	12	—	—	—	—
T42Ald	6	75	19	—	—	—	—
T43Ald	7	19	71	3	—	—	—
T44Ald	12	49	39	—	—	—	—
T45Ald	3	46	48	3	—	—	—

relative sensitivity of various germ-cell stages, very good agreement is found[58-60] with the results of experiments in which genetic damage is measured by dominant lethality or sex chromosome aberrations (Table 4) The frequency of translocated offspring after irradiation of spermatids is about three times greater than that after irradiation of mature spermatozoa and about ten times greater than that after irradiation of the most mature spermatocytes. On the other hand, very few translocated F_1 animals are obtained from irradiated females. In the translocated male offspring, the frequency of translocation configurations varies from 100 to 12% (Table 5), and the same translocation can give different types of configurations. It is noteworthy that although all translocated offspring from irradiated spermatogonia and most of those from postmeiotic irradiated stages are semisterile, a few autosomal translocations causing male sterility can sometimes also be detected after irradiation of postmeiotic male germ cells[58,61] or after irradiation of female mice.[62] Searle,[63] analyzing data of Léonard and Deknudt,[58] has shown that in the males with few or no spermatozoa the frequency of chain configurations was almost twice that of males which had no obvious shortage of spermatozoa (Table 6). Searle[63] postulated that translocations inducing sterility cannot be recovered from spermatogonia because they are selectively eliminated during spermatogonial mitoses.

IV. EXPERIMENTAL RESULTS WITH CHEMICAL MUTAGENS

Experiments with chemical mutagens are still relatively scarce, and careful analysis of the technical procedures that have been used for testing of chemicals indicates that very few experimental investigations have been performed according to standardized methods.

A. Review of Results

1. Spermatocyte Test on Treated Males

Often, the time interval between treatment and observation chosen for the spermatocyte test on treated males was too short. Thus many authors were able only to score some unstable aberrations in spermatogonia or to detect the anomalies induced in spermatocytes. Furthermore, some of the anomalies claimed to be the result of the treatment might even be considered a normal process under the conditions employed. Apparently contradictory results have even been sometimes obtained in different experiments with the same chemical, since the rate of trans-

TABLE 6. Types of Translocation Configuration in Spermatocytes of Presumptively Fertile and Sterile Males after Paternal Postmeiotic Radiation[a]

Number of spermatozoa	Bivalents[b]	Rings[b]	Chains[b]	Total
Normal	519 (17.3)	1366 (45.5)	1115 (37.2)	3000
Lower nil	90 (17.5)	55 (10.7)	368 (71.7)	513

[a]Data of Léonard and Deknudt[58] analyzed and summarized by Searle.[5]
[b]Percentages are in parentheses.

locations observed at the spermatocyte stage and the yield of F_1 translocated sons from treated spermatogonia displayed discrepancies difficult to explain.

   a. Translocations (Table 7). Some rare reciprocal translocations have been observed in the spermatocytes of animals treated with TEM[64] or NEU,[65] but they are more frequent (17.7%) after treatment with TEPA.[66] The most important difference between the results with TEPA and those with ionizing radiations occurred in the ratio between the ring and chain configurations, since 78% of the chromosome rearrangements were in the form of chain quadrivalents. Hydroxylamine[65] and methotrexate[67] are capable of inducing translocations in premeiotic male germs, but negative results have been obtained with all other chemical compounds tested up to now.

   b. X-Y Univalents. An increase in the frequency of X-Y univalents has been reported after treatment with trenimon or cytoxan.[39,68,69] It is noteworthy that these studies have been performed according to a technique[39] involving immersion of the tissue in a hypotonic solution of 1.0% sodium citrate and 0.005% hyaluronidase for 20 min. On the other hand, no increase in frequency of X-Y univalents has ever been observed, for any other chemical compounds, in experiments performed with the technique of Evans et al.,[34] which requires hypotonic treatment in 1% citrate solution for only 10–12 min.

2. F_1 Translocation Test

   Seven chemical compounds (TEM,[2,3] TEPA,[81] nitrogen mustard,[3] MMS,[82] EMS,[83] caffeine,[84] and aflatoxins[79] have been tested for the production of F_1 translocated offspring according to the standard methodology (Table 7). Positive results were obtained with TEM, TEPA, nitrogen mustard, MMS, and EMS, whereas caffeine and aflatoxins failed to induce translocations in the progeny.

   The experiments performed with TEM[2,3,85] and EMS[85] demonstrate that treatment of spermatids with these compounds yields more

TABLE 7. Summary of Experiments on Translocation Induction

Chemical	Spermato-cyte test on treated males	Reference	F_1 trans-location test	Reference
TEM	+	Cattanach and Williams [64]	+	Cattanach [2,3]
Trenimon	−	Schleiermacher [39,68,69]	+?	Datta et al. [80]
TEPA	+	Zudova and Sram [66]	+	Sram et al. [81]
Nitrogen mustard			+	Falconer et al. [1]
Cytoxan	−	Schleiermacher [39,68,69]	+?	Datta et al. [80]
NEU	+?	Ramaiya [65]		
MNNG	−	Léonard et al. [70]		
MMS			+	Léonard and Schröder [82]
EMS	−	Léonard et al. [70]	+	Cattanach et al. [83]
		Cattanach and Williams [64]		
Mitomycin-C	−	Gilliavod and Léonard [71]		
Trypaflavin	−?	Baldermann et al [72]		
EB	−	Léonard et al. [70]		
LSD-25	−	Jagiello and Polani [73]		
Hydroxyla-mine	+?	Ramaiya [65]		
Methotrexate	+?	Haas [67]		
Caffeine	−	Adler [74,75]	−	Cattanach [84]
		Adler and Röhborn [76]		
Cyclamates	−	Cattanach and Pollard [77]		
		Léonard and Linden [78]		
Aflatoxins	−	Léonard and Linden [79]	−	Léonard and Linden [79]

translocations and more dominant lethals than does treatment of spermatozoa. Therefore, the stages of spermatogenesis most sensitive to damage by ionizing radiation also appear to be the ones most sensitive to damage by chemicals.

In the experiments with TEM [2,3] and EMS, [83] the frequency of occurrence of sterile animals closely parallels that of semisterile males. F_1 male sterility was also reported by Falconer et al. [1] following treatment of males with nitrogen mustard. The high frequency of completely sterile males in the F_1 progeny of chemically treated animals contrasts with the results obtained with X-rays [58] and, as pointed out by Cattanach et al., [83] may suggest that "chemical mutagens tend to break chromosomes in different regions and/or cause different types of rearrangements from those induced by X-rays."

The yield of dominant lethals parallels the frequency of translocated F_1 offspring after treatment with TEM [2,3,85] and EMS, [83] and this suggests [83,85] that chromosome breakage is the basis for the chemically induced dominant lethality as well as for the X-ray-induced dominant lethality.

B. Detailed Results

This review will not be restricted to results obtained using standard conditions but will also deal with other pertinent data on reproductive cells. Although some of these experiments are not complete, they may nevertheless yield useful information on the potential ability of the substances to produce translocations. The different chemical compounds have been classified according to Fishbein et al. [86] into the following:

1. Alkylating agents: TEM, trenimon, TEPA, nitrogen mustard, cyclophosphamide, NEU, MNNG, DEB, DES, MMS, EMS, myleran.
2. Base analogues: 5-BUdR.
3. Antibiotics: mitomycin-C.
4. Drugs and miscellaneous mutagens: trypaflavin, EB, LSD-25, hydroxylamine, progesterone, methotrexate.
5. Food additives or components: caffeine, cyclamates.
6. Food contaminants: aflatoxins.

1. Alkylating Agents

The alkylating agents constitute the largest category studied and exhibit a diversity of pharmacological properties. Many alkylating agents have been found to exert a mutagenic effect on a great variety of organisms. Experiments *in vitro* indicate that mainly chromatid breaks are produced, but the occurrence of dominant lethals and translocations also proves the ability of alkylating agents to cause chromosome-type aberrations.

a. Triethylenemelamine [*2,4,6-Tris(1-aziridinyl)-s-triazine; Tretamine; TEM*]. Intraperitoneal injection of TEM induces chromosome abnormalities in cultured leukocytes [87] and appears highly effective in breaking rodent chromosomes, as indicated by the induction of dominant lethals in postmeiotic germ cells of the rat [88–90] and of the mouse. [2,3,91–93]

A low incidence (0.55%) of translocations and chromosome fragments has been reported in mice killed 12 weeks after a dose of 0.5 or 2.0 mg TEM per kilograms of body weight. [64] Some chromosome fragments and bridges have been observed in anaphases I and II of male

TABLE 8. Analysis of the F_1 Generation from Males Given TEM[3]

Days conceived after injection of fathers	Number of F_1 males	Number semisterile	Number sterile	Number translocations	Percent translocations
1– 3	111	4	4	5	4.51
10–13	74	10	12	11	14.86

mice given 6 mg TEM/kg and killed 20 hr or 2, 4, 5, 7, 10, 20, or 30 days after injection.[94]

Offspring carrying translocations have been observed by Cattanach[2,3] among the progeny of male mice given intraperitoneal injections of TEM. In a first experiment,[2] the animals received two intraperitoneal injections of solutions of TEM (0.64 or 0.80 mg TEM/kg body weight) and sired about 30% translocated offspring. In more extensive experiments,[3] male mice were given an intraperitoneal injection of a solution of 0.2 mg TEM/kg made up with 0.9% sodium chloride, and 4.51% of the males sired on the first 3 days and 14.86% of the males sired on the tenth to thirteenth days were found to be heterozygous for a translocation. These important results have been discussed previously (Table 8).

b. Trenimon (2, 3, 5-Trisethyleneimino-1, 4-benzoquinone). Achromatic lesions, chromatid breaks, and chromatid translocations were observed after treatment with trenimon in human leukocytes *in vitro*[95] and in the bone marrow of Chinese hamsters treated *in vitro*[96] as well as in cell culture systems.[97] Trenimon also induces dominant lethals in male mice,[98] the rate in spermatocytes and spermatogonia being lower than in spermatids or spermatozoa.

An increase in autosomal and X-Y univalent frequency has been found[39,68,69] in male mice injected intraperitoneally with 0.0625–0.3 mg trenimon/kg body weight and sacrificed 6 hr to 31 days after treatment.

Datta *et al.*[80] analyzed the mitoses in embryonic liver from 39 F_1 offspring from spermatozoa or spermatids of mice given 0.125 mg trenimon/kg intraperitoneally. They detected two animals carrying translocations, as shown by the presence of a very large acrocentric chromosome. One should expect, therefore, that an examination of meiotic chromosomes would confirm the induction of translocations by treatment of postmeiotic male germ cells.

c. TEPA [Tris(1-aziridinyl)phosphine Oxide; Triethylenephosphoramide; Aphoxide; APO]. TEPA induces gaps and chromatid breaks in human cells *in vitro*[99] and dominant lethals in postmeiotic male germ cells.[93,100]

This chemical compound gave 17.7% cells carrying translocations in animals given 40 mg TEPA/kg intraperitoneally and killed 10 weeks after treatment. As pointed out previously, the most important feature was the high incidence of chain quadrivalents.[66] In this experiment, the meiotic preparations were made by the method of Meredith.[43]

This chemical compound has also been reported[81] to induce a very high incidence of translocated F_1 offspring after treatment of spermatogonial stages as well as after treatment of postmeiotic male germ cells.[101]

d. *Nitrogen Mustard* [*Methylbis(β-chloroethyl)amine Hydrochloride*]. Fertility tests indicate[91] that nitrogen mustard can produce dominant lethals in mice.

Analyzing the progeny of 12 males treated intraperitoneally with nitrogen mustard, Falconer *et al.*[1] have observed one heterozygote for a visible recessive, two completely sterile males, and three suspected semisteriles. One of the semisterile males was analyzed cytologically and was found to carry two independent translocations.

e. *Cyclophosphamide* [N,N-*Bis(2-chloroethyl)*-N',O-*propylenephosphoric Acid Ester Diamide; Endoxan; Cytoxan*]. Cyclophosphamide produces mainly chromatid-type aberrations in small mammals and in human lymphocytes.[102–108] It also induces dominant lethals in premeiotic and in postmeiotic male germ cells.[98,109,110]

Some increase in the yield of autosomal and X-Y univalents has been observed in animals given 200 mg cyclophosphamide/kg and killed 4–22 days after treatment.[39,68,69]

Possibly, the examination of meiotic chromosomes from F_1 sons might have provided evidence for the induction of translocations in postmeiotic male germ cells, since two translocated offspring have been observed by Datta *et al.*,[80] who examined embryonic liver mitoses of 34 F_1 sons from spermatozoa or spermatids of animals given 210 mg cyclophosphamide/kg.

f. *NEU* (N-*Ethyl*-N-*nitrosurea*). Chromosome aberrations in the male germ cells of mice treated intraperitoneally with 50 or 100 mg NEU/kg are reported 21, 123, or 312 hr after treatment.[65] However, the interval of time between treatment and observation is too short to conclude that the anomalies were already induced in the spermatogonia stage.

g. N-*Methyl*-N'-*nitro*-N-*nitrosoguanidine (MNNG)*. MNNG induces chromatid aberrations in human cells *in vitro*[111] but has been found ineffective in the induction of dominant lethal mutations in postmeiotic male germ cells of the mouse.[112]

No chromosome rearrangements have been observed in BALB/c mice injected intraperitoneally with 48 mg MNNG/kg and killed 120 days after treatment.[70]

h. dl-*Diepoxybutane (Butadiene Diepoxide; DEB)*. Chromosome fragments and bridges in anaphase I and II were observed in animals given

TABLE 9. Induction of Translocation in F₁ Females and Males by EMS[a]

Day conceived	Females		Males				Total percent translocations
	Number tested	Number of translocations	Number tested	Number semisterile	Number sterile	Number of translocations	
1	7	0	9	0	0	0	0 ⎫ 7–9
2	9	2	5	0	0	0	14 ⎭
3	21	3 (+ 1)	19	0	2	0	7.5–10
4	20	1	30	3	3	7	16 ⎫ 15–16
5	13	1 (+ 1)	17	2	1	3	13–17 ⎭
6	4	2	3	0	2	3	57 ⎫ 50
7	3	2	4	0	1	2	43 ⎭
8	3	2	1	0	1	1	50
9	5	1 (+ 1)	9	3	0	0	29–36 ⎫ 28–38
10	4	(1)	7	2	1	3	27–36 ⎭
Total	89	14 (+ 4)	104	10	10	19	17–19

[a] Numbers in parentheses indicate animals tested genetically but not cytologically.

5 mg DEB/kg and killed 2, 3, 4, 5, 7, 10, 20, or 30 days after injection. [94]

i. Diethyl Sulfate (DES). No translocations have been detected among the 2000 spermatocytes scored in ten mice killed after intraperitoneal injection of 177 mg DES/kg. [70]

j. Methyl Methanesulfonate (MMS). MMS induces dominant lethals in postmeiotic male germ cells. [93,112–113] Several translocated males have been observed among the F_1 progeny of male mice given intraperitoneal injection of 50 mg MMS/kg. [82]

k. Ethyl Methanesulfonate (EMS). EMS produces dominant lethals in postmeiotic male germ cells. [83,112,114] No translocations have been scored among 2000 cells examined from ten animals given intraperitoneal injection of 336 mg EMS/kg. [70] Similar results are reported with 240 mg EMS/kg. [64]

The results of Cattanach *et al.* [83] with the F_1 translocation test have already been discussed. As shown in Table 9, this compound induces a very high frequency of F_1 translocated male and female offspring: the incidence of translocation heterozygotes was 7–9% for the animals sired the first 3 days, 15–16% for the animals sired on the fourth to the fifth days, 28–38% for the animals sired on the eighth to the tenth days, and up to 50% for the animals sired on the sixth to the seventh days. Comparison of the EMS [83] and TEM experiments [4,5] demonstrates that "at equimolar doses TEM is by far the more effective agent." [83]

l. Myleran [1, 4-Di(methanesulfonyloxy)butane; Busulfan]. Myleran causes chromosome aberrations in human lymphocytes [115] and dominant lethals in male rats. [114]

It produces chromosome fragments and bridges in anaphases I and II of male mice given 16 mg myleran/kg and killed 2, 3, 4, 5, 7, 10, 20, or 30 days after treatment. [94]

2. Base Analogues

a. 5-Bromodeoxyuridine (5-BUdR). 5-BUdR produces many chromatid breaks in murine cells [116] but only a few chromosome breaks in human cell cultures [117,118] and no dominant lethals in male mice. [93]

Fragmentation and stickiness have been observed in spermatogonial metaphases and different stages of meiotic prophase of Chinese hamsters given 1 μg 5-BUdR and killed 3, 6, 12, 24, 48, or 60 hr after treatment. [119]

3. Antibiotics

a. Mitomycin-C (MC). Mitomycin-C produces chromatid-type aberrations in leukocyte or fibroblast cultures [120–123] and dominant lethality in early spermatids and in spermatocytes. [124]

Fifty days after intraperitoneal injection of 5 mg mitomycin/kg, testis weight decreased significantly from a mean of 250 ± 5 mg for the controls to 102 ± 12 mg for the treated males. No chromosome rearrangements were observed, and the rate of X-Y univalents as well as the incidence of polyploidy appeared to be normal. [71]

4. Drugs and Miscellaneous Mutagens

a. *Trypaflavin (an Acridine Derivative).* An increase in frequency of gaps has been observed in HeLa cells and lymphocyte cultures [125–127] after exposure to a dose of 10^{-6} mol trypaflavin/liter. An oral dose of 50 mg/kg induced dominant lethals in pre- and postmeiotic male germ cells. [72]

The testes of a few males examined [72] by the technique of Schleiermacher [39] showed a slight increase in the yield of univalents, but the number of spermatocytes scored is too small to allow any conclusion.

b. *Ethidium Bromide (Homidium Bromide; 2, 7-Diamino-9-phenyl-10-ethylphenanthridinium Bromide; EB).* No chromosome rearrangements have been found among 2000 spermatocytes scored in ten male mice killed 120 days after an intraperitoneal injection of a dose of 3.9 mg EB/kg. [70]

c. *LSD-25 (Lysergic Acid Diethylamide).* Chromosome damage induced *in vitro* by LSD is generally of the chromatid type [128] and occurs only at a concentration of drug and a duration of exposure which would be impossible to obtain in humans with a reasonable dose.

Some gaps, breaks, fragments, secondary constrictions, and metacentric chromosomes have been described in the germ cells of animals given 25 μg LSD/kg [129] and killed a few days after treatment. However, no evidence for structural or numerical changes was found by Jagiello and Polani [73] after a single subcutaneous dose of LSD-25 (1 mg/kg) or in chronic experiments with daily doses of 0.0033–0.18 mg/kg for 31 days. In the acute experiments, animals were killed 12–66 hr, and in the chronic ones 1–23 days after the last injection. These conclusions are confirmed by the negative results obtained on primate germ cells by Hulten et al. [131] 6 months after the last injection of a very large amount of LSD and other drugs.

d. *Hydroxylamine.* Hydroxylamine produces chromosome aberrations in human cells, [118] cultured Chinese hamster cells, [132] and mouse embryo cells. [133]

It is possible that hydroxylamine can also induce fragments and chromosome rearrangements in spermatogonia. [65]

e. *Progesterone.* Slight alterations such as sticky degeneration and improper spreading with clumping of the meiotic chromosomes are reported in a dog given progesterone at a constant rate directly into the spermatic artery and examined at the end of perfusion. [134]

f. *Methotrexate (Amethopterin).* The folic acid antagonist methotrexate has been claimed to increase aneuploidy[135] in bone marrow and lymphocyte cultures from patients receiving total doses ranging between 570 and 2735 mg.

After treatment of spermatogonia with 200–500 mg methotrexate/kg, terminal deletions and reciprocal translocations have been observed by Haas.[67]

5. Food Additives or Components

a. *Caffeine (1, 3, 7-Trimethylxanthine)*. The frequency of chromosome breakage produced by caffeine in human cells in culture,[136–140] in mouse tissue cultures,[141] and in Chinese hamster cells[142] tends to increase linearly with the dose, but the yield of translocations is rather low. Caffeine has been shown to be ineffective in producing dominant lethals by Lyon *et al.*,[143] Epstein,[144] and Epstein *et al.*,[145] and the positive results obtained by Kulman *et al.*[146] are not significant, since the number of females examined was very small.

The spermatocyte test on male mice given caffeine for a long period in their drinking water[74,76] or treated with a single dose of 0.2 g/kg or with daily doses of 0.25 g/kg for 3 weeks gave negative results.[75]

No translocation could be found cytologically[84] in the 201 F_1 sons sired by JU males given 0.3% solution of caffeine as their drinking water for 3 months.

b. *Cyclamates (Cyclamic Acid; Cyclohexanesulfamic Acid; and Related Compounds)*. Sodium and calcium cyclamates and related compounds have been claimed to increase chromosome damage in human cells *in vitro*[147–148] and in somatic cells of small mammals,[149–150] but negative results are reported by Brewen *et al.*[151] in human leukocytes and in bone marrow cells from Chinese hamsters. Petersen *et al.*[152] reported dominant lethals in postmeiotic male germ cells, but Cattanach and Pollard[77] did not confirm this.

Five daily intraperitoneal injections of cyclohexylamine did not produce translocations in the animals killed 8 or more weeks later.[77] Negative results were also obtained in mice given solutions of 2.7, 5.4, or 10.8 g sodium cyclamate/liter of water as their drink for periods of 30, 60, or 150 days.[78]

6. Food Contaminants

a. *Aflatoxins*. "Aflatoxins" is a collective term for eight compounds of related molecular configuration (i.e., aflatoxins B1, B2, B2a, G1, G2, G2a, M1, and M2) which are produced by some strains of fungi, e.g., *Aspergillus parasiticus* and *Penicillium puberulum*. Aflatoxin B1 inhibits the mitotic process in human embryonic lung cells [153] and produces chromosome aberrations in human leukocytes *in vitro.* [154,155]

A single intraperitoneal injection of 5 mg aflatoxin B1 in 10% saline solution of dimethylsulfoxide (1 ml dimethylsulfoxide plus 9 ml 9% NaCl) does not produce chromosome rearrangement in the treated animals or in their offspring. [79]

C. Evaluation of Results and Prospects of the Experimental Approach

In the studies *in vitro,* lymphocytes are exposed to doses of chemicals higher than those used normally in the intact organism. Furthermore, the *in vivo* effect of a chemical depends on its penetration and distribution in the body and can be modified by metabolic degradation and by the rate of excretion. Therefore, compounds mutagenic *in vitro* must not necessarily be so *in vivo,* and the hazard to man can be properly evaluated only on the basis of experiments on living mammals.

As pointed out previously, the spermatocyte test on treated males and the F_1 translocation test have many advantages over the specific-loci method and the dominant lethal test. Our analysis of the experimental data shows that among the compounds studied according to the standard methodology of the spermatocyte test, TEPA and to some extent TEM appear to be the only ones able to induce chromosome rearrangements in spermatogonia. Positive responses have been reported more frequently with the F_1 translocation test on postmeiotic male germ cells. Six compounds have been tested so far by both methods: aflatoxin B1 and caffeine give negative results for pre- and postmeiotic germ cells, TEM and TEPA induce chromosome rearrangements in pre- and postmeiotic male germ cells, whereas MMS and EMS produce translocations only in postmeiotic male germ cells and not in premeiotic ones (Table 7). These results also show that one cannot predict whether a chemical that produces effects in postmeiotic stages will do so in premeiotic ones. Therefore, in view of the fact that spermatogonia rather than postmeiotic germ cells are the population at major risk, the spermatocyte test on treated males must be considered as the most suitable and the cheapest method.

V. ACKNOWLEDGMENTS

I am very grateful to Dr. C. Auerbach, Dr. B. M. Cattanach, and Dr. A. G. Searle, who carefully reviewed the manuscript. I am indebted to Dr. G. Gerber for helpful discussion and for revising the English text. I also thank the editors and authors for permission to reproduce Tables 4, 5, 6, and 8.

VI. REFERENCES

1. D. S. Falconer, B. M. Slyzinski, and C. Auerbach, Genetical effects of nitrogen mustard in the house mouse, *J. Genet. 51*, 81–88 (1952).
2. B. M. Cattanach, Induction of translocations in mice by triethylenemelamine, *Nature 180*, 1364–1365 (1957).
3. B. M. Cattanach, The sensitivity of the mouse testis to the mutagenic action of triethylenemelamine, *Z. Indukt. Abstammungs-Vererbungsl. 90*, 1–6 (1959).
4. A. Léonard, Radiation-induced translocations in spermatogonia of mice, *Mutation Res. 11*, 71–88 (1971).
5. A. G. Searle, Chromosome damage and risk assessment, in "Proceedings IV International Congress of Human Genetics" (J. de Grouchy, F. J. G. Ebling, and I. W. Henderson, eds.) pp. 58–66, Excerpta Medica, Amsterdam (1972).
6. E. F. Oakberg, Duration of spermatogenesis in the mouse and timing of stages of the cycle of the seminiferous epithelium, *Am. J. Anat. 99*, 507–516 (1956).
7. Y. Clermont and E. Bustos-Obregon, Reexamination of spermatogonial renewal in the rat by means of seminiferous tubules mounted "in toto," *Am. J. Anat. 122*, 237–248 (1968).
8. E. F. Oakberg, A new concept of spermatogonial stem-cell renewal in the mouse and its relationship to genetic effects, *Mutation Res. 11*, 1–7 (1971).
9. E. F. Oakberg, Spermatogonial stem-cell renewal in the mouse, *Anat. Rec. 169*, 515–532 (1971).
10. C. Huckins, The spermatogonial stem cell population in adult rats, 1. Their morphology, proliferation and maturation, *Anat. Rec. 169*, 533–558 (1971).
11. E. F. Oakberg, Duration of spermatogenesis in the mouse, *Nature 180*, 1137–1139 (1957).
12. J. L. Sirlin and R. G. Edwards, Duration of spermatogenesis in the mouse, *Nature 180*, 1137–1138 (1957).
13. E. F. Oakberg and R. L. DiMinno, X-ray sensitivity of primary spermatocytes of the mouse, *Internat. J. Radiation Biol. 2*, 196–209 (1960).
14. S. Kofman-Alfaro and A. C. Chandley, Meiosis in the male mouse. An autoradiographic investigation, *Chromosoma (Berlin) 31*, 404–420 (1970).
15. A. Léonard and Gh. Deknudt, Persistence of chromosome rearrangements induced in male mice by X-irradiation of pre-meiotic germ cells, *Mutation Res. 9*, 127–133 (1970).
16. A. Léonard and Gh. Deknudt, Réarrangements chromosomiques induits par les rayons X chez les souris de race AKR/T1Ald, *Experientia 22*, 715 (1966).
17. E. P. Evans, M. F. Lyon, and M. Daglish, A mouse translocation giving a metacentric marker chromosome, *Cytogenetics 3*, 159–166 (1967).
18. B. J. White and J. H. Tjio, A mouse translocation with 38 and 39 chromosomes but normal N.F., *Hereditas 58*, 284–296 (1967).
19. C. E. Ford, Meiosis in mammals, in "Comparative Mammalian Cytogenetics" (K. Benirschke, ed.) pp. 91–106, Springer-Verlag, Berlin, Heidelberg, New York (1969).
20. C.-C. Lin, W. S. Tsuchida, and S. A. Morris, Spontaneous meiotic chromosome abnormalities in male mice (*Mus musculus*), *Can. J. Genet. Cytol. 13*, 95–100 (1971).
21. C. E. Ford and E. P. Evans. Origin of apparent polyploid spermatocytes in the mouse, *Nature 230*, 389–390 (1971).
22. A. Léonard, Unpublished results (1972).
23. A. G. Searle, C. V. Beechey, E. P. Evans, C. E. Ford, and D. G. Papworth, Studies

on the induction of translocations in mouse spermatogonia, IV. Effects of acute gamma-irradiation, *Mutation Res. 12,* 411–416 (1971).

24.   P. C. Koller and C. A. Auerbach, Chromosome breakage and sterility in the mouse, *Nature 148,* 501–502 (1941).

25.   P. C. Koller, Segmental interchange in mice, *Genetics 29,* 247–263 (1944).

26.   C. Auerbach, Unpublished data, Conclusions and comments of a workshop held in U.S.A. (1971).

27.   J. Lejeune, B. Dutrillaux, and J. de Grouchy, Reciprocal translocations in human populations, a preliminary analysis, *in* "Human Population Cytogenetics" (P. A. Jacobs, W. H. Price, and P. Law, eds.) pp. 82–87, University Press, Edinburgh (1970).

28.   C. E. Ford and J. L. Hamerton, A colchicine hypotonic citrate squash sequence for mammalian chromosomes, *Stain Technol. 31,* 247–251 (1956).

29.   S. Ohno, W. D. Kaplan, and R. Kinosita, Heterochromatic regions and nucleolus organizers in chromosomes of the mouse, *Mus musculus, Exptl. Cell Res. 13,* 358–364 (1957).

30.   M. C. Bunker, A technique for staining chromosomes of the mouse with Sudan Black B., *Can. J. Genet. Cytol. 3,* 355–360 (1961).

31.   C. E. Ford, Methods in human cytogenetics, *in* "Methodology in Human Genetics" (W. J. Burdette, ed.) pp. 227–259, San Francisco, Holden–Day (1962).

32.   W. J. Welshons, B. H. Gibson, and B. J. Scandlyn, Slide processing for the examination of male mammalian meioti chromosomes, *Stain Technol.37,* 1–5 (1962).

33.   H. Gardner and H. Punett, An improved squash technique for human male meiotic chromosomes: Softening and concentration of cells; mounting in Hoyer's medium, *Stain Technol. 39,* 245–248 (1964).

34.   E. P. Evans, G. Breckon, and C. E. Ford, An air-drying method for meiotic preparations from mammalian testes, *Cytogenetics 3,* 289–294 (1964).

35.   K. H. Rothfels and L. Siminovitch, An air-drying technique for flattening chromosomes in mammalian cells grown *in vitro, Stain Technol. 33,* 73–77 (1958).

36.   J. H. Tjio and T. T. Puck, Genetics of somatic mammalian cells. II. Chromosomal constitution of cells in tissue culture, *J. Exptl. Med. 108,* 259–268 (1958).

37.   C. E. Ford, The use of chromosome markers, *in* "Tissue Grafting and Radiation" (H. S. Micklem and J. F. Loutit, eds.) pp. 197–206, Academic Press, New York (1966).

38.   K. Benirschke and L. E. Brownhill, Heterosexual cells in testes of chimeric marmoset monkeys, *Cytogenetics 2,* 331–341 (1963).

39.   E. Schleiermacher, Über den Einfluss von Trenimon und Endoxan auf die Meiose der männlichen Maus, *Humangenetik 3,* 134–155 (1966).

40.   D. L. Williams, A. A. Hagen, J. W. Runyan, and D. A. Lofferty, A method for the differentiation of male meiotic chromosome stages, *J. Hered. 62,* 17–22 (1971).

41.   C. E. Ford and E. P. Evans, Meiotic preparations from mammalian testes, *in* "Comparative Mammalian Cytogenetics (K. Benirschke, ed.) pp. 461–464, Springer-Verlag, Berlin, Heidelberg, New York (1969).

42.   E. M. Eicher, An air-drying procedure for mammalian male meiotic chromosomes, following softening in gluconic acid and cell separation by an ethanol–acetic mixture, *Stain Technol. 41,* 317–321 (1966).

43.   R. Meredith, A simple method for preparing meiotic chromosomes from mammalian testis, *Chromosoma 26,* 254–258 (1969).

44.   A. P. Dyban, An improved air-drying method for meiotic and mitotic chromosome preparation from mammalian testes, *Tsitologiva 12,* 687–690 (1970).

45.   S. H. Hoo and C. A. Bowles, A unique air-drying method for preparing metaphase

chromosomes from the spermatogonial cells of rats and mice, *Mutation Res. 13*, 85–88 (1971).

46. A. Léonard and Gh. Deknudt, Meiotic chromosome rearrangements induced in mice by irradiation of spermatogonial stages, *Can. J. Genet. Cytol. 8*, 520–527 (1966).

47. A. Léonard and Gh. Deknudt, Relation between the X-ray dose and the rate of chromosome rearrangements in spermatogonia of mice, *Radiation Res. 32*, 35–41 (1967).

48. A. Léonard and Gh. Deknudt, Chromosome rearrangements after low X-ray doses given to spermatogonia of mice, *Can. J. Genet. Cytol. 10*, 119–124 (1968).

49. A. Léonard and Gh. Deknudt, Dose-response relationship for translocations induced by 2-irradiation in spermatogonia of mice, *Radiation Res. 40*, 276–284 (1969).

50. A. Léonard and Gh. Deknudt, The rate of translocations induced in spermatogonia of mice by two X-irradiation exposures separated by varying time intervals, *Radiation Res. 45*, 72–79 (1971).

51. N. V. Savkovic and M. F. Lyon, Dose response curve for X-ray induced translocations in mouse spermatogonia. I. Single doses, *Mutation Res. 9*, 407–409 (1970).

52. E. P. Evans, C. E. Ford, A. G. Searle, and B. J. West, Studies on the induction of translocations in mouse spermatogonia. II. Effects of X-irradiation, *Mutation Res. 9*, 501–506 (1970).

53. S. Muramatsu, W. Nakamura, and H. Eto, Radiation-induced translocations in mouse spermatogonia, *Japan. J. Genet. 46*, 281–283 (1971).

54. G. B. Gerber and A. Léonard, Influence of selection, non-uniform cell population and repair in dose effect curves of genetic effects, *Mutation Res. 12*, 175–182 (1971).

55. A. G. Searle, E. P. Evans, C. E. Ford, and B. J. West, Studies on the induction of translocations in mouse spermatogonia. I. The effects of dose rate, *Mutation Res. 6*, 427–436 (1968).

56. A. G. Searle, E. P. Evans, and C. E. Ford, A comparison of cytological and genetical observations on the yield of major chromosome rearrangements following irradiation of mouse spermatogonia, *Ann. Human Genet. 29*, 11 (1965).

57. C. E. Ford, A. G. Searle, E. P. Evans, and B. J. West, Differential transmission of translocations induced in spermatogonia of mice by irradiation, *Cytogenetics 8*, 447–470 (1969).

58. A. Léonard and Gh. Deknudt, The sensitivity of various germ-cell stages of the male mouse to radiation induced translocations, *Can. J. Genet. Cytol. 10*, 495–507 (1968).

59. A. Léonard, Differential radiosensitivity of germ-cells of the male mouse, *Can. J. Genet. Cytol. 7*, 400–405 (1965).

60. L. B. Russell and C. L. Saylors, The relative sensitivity of various germ cell stages of the mouse to radiation-induced non-disjunction, chromosome losses and deficiencies, *in* "Repair From Genetic Damage" (F. H. Sobels, ed.) pp. 313–340, Pergamon Press, Oxford (1963).

61. M. F. Lyon and R. Meredith, Autosomal translocations causing male sterility and viable aneuploidy in the mouse, *Cytogenetics 5*, 335–354 (1966).

62. A. G. Searle and C. V. Beechey, Unpublished information, cited in Ref. 2 (1971).

63. A. G. Searle, Symposium on mammalian radiation genetics, Summary and synthesis (July 7–9, 1970), *Mutation Res. 11*, 133–147 (1971).

64. B. M. Cattanach and C. E. Williams, A search for chromosome aberrations induced in mouse spermatogonia by chemical mutagens, *Mutation Res. 13*, 371–375 (1971).

65. L. K. Ramaiya, The cytogenetic effect of *N*-nitrosoethyl-urea, hydroxylamine and X-rays on the germ-cells of male mice, *Genetika 5 (2)*, 74–86 (1969).

66. Z. Zudova and R. J. Sram, Effects of TEPA on the induction of chromosome rearrangements in spermatogonia of mice, *EMS Newsletter 4*, 41 (1971).

67. E. Haas, Die Wirkung von Methotrexat auf die Spermatogenese der Maus, Dissertation, Heidelberg (1970).

68. E. Schleiermacher, Über den Einfuss von Trenimon und Endoxan auf die Meiose der männerlichen Maus. I. Methodik der Präparation und Analyse meiotischer Teilungen, *Human genetik 3*, 127–133 (1966).

69. E. Schleiermacher, The activity of alkylating agents. II. Histological and cytogenetic finding in spermatogenesis, *in* "Chemical Mutagenesis in Mammals and Man" (F. Vogel and G. Röhrborn, eds.) pp. 317–341, Springer-Verlag, Berlin, Heidelberg, New York (1970).

70. A. Léonard, Gh. Deknudt, and G. Linden, Failure to detect meiotic chromosome rearrangement in male mice given chemical mutagens, *Mutation Res. 13*, 89–92 (1971).

71. N. Gilliavod and A. Léonard, Tests for mutagenic effects of chemicals in mice. I. Effects of mitomycin C on spermatogonia, *Mutation Res. 13*, 274–275 (1971).

72. K. H. Baldermann, G. Röhrborn, and T. M. Schroeder, Mutagenitätsunter Suchungen mit Trypaflavin und Hexamethylentetramin am Sänger *in vivo* und *in vitro, Humangenetik 4*, 112–126 (1967).

73. G. Jagiello and P. E. Polani, Mouse germ cells and LSD-25, *Cytogenetics 8*, 136–147 (1969).

74. I. D. Adler, Cytogenetic investigations of mutagenic action of caffeine in premeiotic spermiogenesis in mice, *Humangenetik 3*, 82–83 (1966).

75. I. D. Adler, The problem of caffeine mutagenicity, *in* "Chemical Mutagenesis in Mammals and Man" (F. Vogel and G. Röhrborn, eds.) pp. 383–403, Springer-Verlag, Berlin, Heidelberg, New York (1970).

76. I. D. Adler and G. Röhrborn, Cytogenetic investigation of meitotic chromosomes of male mice after chronic caffeine treatment, *Humangenetik 8*, 81–85 (1969).

77. B. M. Cattanach and C. E. Pollard, Mutagenicity tests with cyclohexylamine in the mouse, *Mutation Res. 12*, 472–474 (1971).

78. A. Léonard and G. Linden, Observations sur les propriétés mutagéniques des cyclamates chez les mammifères, *C. R. Soc. Biol. 166*, 468–470 (1972).

79   A. Léonard and G. Linden, Mutagenicity studies with aflatoxins in the mouse, In preparation (1973).

80. P. K. Datta, H. Frigger, and E. Schleiermacher, The effect of chemical mutagens on the mitotic chromosomes of the mouse, *in vivo*, *in* "Chemical Mutagenesis in Mammals and Man" (F. Vogel and G. Röhrborn, eds.) pp. 194–213, Springer-Verlag, Berlin, Heidelberg, New York (1970).

81. R. J. Sram. Z. Zudova, and V. Benes, Induction of translocations in mice by TEPA, *Folia Biol. 16*, 367–368 (1970).

82. A. Léonard and J. H. Schröder, Induction of translocations by the treatment of mouse post-meiotic germ cells with MMS, In preparation (1973).

83. B. M. Cattanach, C. E. Pollard, and J. H. Issacson, Ethyl methanesulfonate–induced chromosome breakage in the mouse, *Mutation Res. 6*, 297–307 (1968).

84. B. M. Cattanach, Genetical effects of caffeine in mice, *Z. Vererbungsl. 93*, 215–219 (1962).

85. B. M. Cattanach and R. G. Edwards, The effects of triethylenemelamine on the fertility of male mice, *Proc. Roy. Soc. Edinburgh 67*, 54–64 (1958).

86. L. Fishbein, W. G. Flamm, and H. L. Falk, "Chemical Mutagens; Environmental Effects on Biological Systems," 364 pp., Academic Press, New York and London (1970).

87. K. E. Hampel and H. Gerhartz, Strukturanomalien der Chromosomen menschlicher Leukozyten *in vitro* durch Triethylenmelamin, *Exptl. Cell Res. 37*, 251 (1965).

88. H. Jackson and M. Bock, Effects of triethylenemelamine on the fertility of rats, *Nature 175*, 1037 (1955).

89. M. Bock and H. Jackson, The action of triethylenemelanine on the fertility of male rats, *Brit. J. Pharmacol. 12*, 6 (1957).

90. A. W. Craig, B. W. Fox, and H. Jackson, Sensitivity of the spermatogenic process in the rat to radiomimetic drugs and X-rays, *Nature 181*, 353 (1958).

91. B. M. Cattanach, A genetical approach to the effects of radiomimetic chemicals on fertility in mice, in "Effects of Ionizing Radiation on the Reproductive System" (W. D. Carlson and F. X. Gassner, eds) pp. 415–426, Pergamon Press, Oxford, London, New York, Paris (1964).

92. A. J. Bateman, The induction of dominant lethal mutations in rats and mice by triethylenemelanine (TEM), *Genet. Res. 1*, 381–392(1960).

93. S. S. Epstein and H. Schafner, Chemical mutagens in the human environment, *Nature 219*, 385–387 (1968).

94. J. Moutschen, Differential sensitivity of mouse spermatogenesis to alkylating agents, *Genetics 46*, 291–299 (1961).

95. G. Obe, Chemische Konstitution und mutagene Wirkung. V. Vergleichende Untersuchung der Wirkung von Athyleniminen auf menschliche Leukozytenchromosomes, *Mutation Res. 6*, 467–471 (1968).

96. W. Schmid, D. T. Arakaki, N. A. Breslau, and J. C. Culberston, Chemical mutagenesis, the Chinese hamster bone marrow as an *in vivo* test system. I. Cytogenetic results on basic aspects of the methodology, obtained with alkylating agents, *Humangenetik 11*, 103–118 (1971).

97. D. T. Arakaki and W. Schmid, Chemical mutagenesis, the Chinese hamster bone marrow as an *in vivo* test system. II. Correlation with *in vitro* results on Chinese hamster fibroblasts and human fibroblasts and lymphocytes, *Humangenetik 11*, 119–131 (1971).

98. G. Röhrborn, The activity of alkylating agents. I. Sensitive mutable stages in spermatogenesis and oogenesis, *in"* Chemical Mutagenesis in Mammals and Man" (F. Vogel and G. Röhrborn, eds.) pp. 294–316, Springer-Verlag, Berlin, Heidelberg, New York (1970).

99. T.–H. Chang and W. Klassen, Comparative effects of tretamine, tepa, apholate and their structural analogues on human chromosomes *in vitro, Chromosoma 24*, 314–323 (1968).

100. R. J. Sram, V. Benes, and Z. Zudova, Induction of dominant lethals in mice by TEPA and HEMPA, *Folia Biol. (Praha) 16*, 407–416 (1970).

101. R. J. Sram, Z. Zudova, V. Benes, and K. Symon, Induction of translocations in mice by TEPA, *EMS Newsletter 3*, 13–14 (1970).

102. F. E. Arrighi, T. C. Hsu, and D. F. Bersagel, Chromosome damage in murine and human cells following cytoxan therapy, *Texas Rep. Biol. Med. 20*, 545–549 (1962).

103. M. Bauchinger and E. Schmid, Cytogenetische Veränderungen in weissen Blutzellen nach Cyclophosphamidtherapie, *Z. Krebsforsch. 72*, 77–87 (1969).

104. E. Schmid and M. Bauchinger, Chromosomenaberrationen in menschlichen Lymphozyten nach Endoxan. Stosstherapie gynäkologischer Tumoren, *Deutsch. Med. Wschr. 93*, 1149–1151 (1968).

105. W. Schmid and G. R. Staiger, Chromosome studies on bone marrow from Chinese hamsters treated with benzodiazepine tranquillizers and cyclophosphamide, *Mutation Res. 7*, 99–108 (1969).

106. M. Vrba, Wirkung von Endoxan auf die Chromosomen von HeLa-Zellen, *Humangenetik 4*, 362–370 (1967).

107. A. Michaelis and R. Rieger, Interaction of chromatid breaks induced by three different radiomimetic compounds, *Nature 199*, 1014–1015 (1961).

108. P. K. Datta and E. Schleiermacher, The effects of cytoxan on the chromosomes of mouse bone marrow, *Mutation Res. 8*, 623–628 (1969).

109. D. Brittinger, Die mutagene Wirkung von Endoxan bei der Maus, *Humangenetik 3*, 156–165 (1966).

110. G. Röhrborn and F. Vogel, A search for dominant mutations in F_1 progeny of male mice treated with Trenimone (triethyleneiminobenzoquinone-1,4), *Human Genet. 7*, 43–50 (1969).

111. F. Kelly and M. Legator, Effect of N-methyl-N'-nitroso-N-nitrosoguanidine on the cell cycle and chromosomes of human embryonic lung cell, *Mutation Res. 10*, 237–246 (1970).

112. U. H. Ehling, R. B. Cumming, and H. V. Malling, Induction of dominant lethal mutations by alkylating agents in male mice, *Mutation Res. 5*, 417–428 (1968).

113. M. Partington and A. J. Bateman, Dominant lethal mutations induced in male mice by methyl methanesulfonate, *Heredity 19*, 191–200 (1964).

114. M. Partington and H. Jackson, The induction of dominant lethal mutations in rats by alkane sulphoric esters, *Genet. Res. 4*, 333–345 (1963).

115. E. Gebhart, Zur Beeinflussung der Wirkung von Myleran auf menschliche Chromosomen durch L-Cystein, *Mutation Res. 7*, 254–257 (1969).

116. T. C. Hsu and C. E. Somers, Effect of 5-bromodeoxyuridine on mammalian chromosomes, *Proc. Natl. Acad. Sci. 47*, 396–403 (1961).

117 M. M. Kaback, E. Saksela, and W. J. Mellmann, The effect of 5-bromodeoxyuridine on human chromosomes, *Exptl. Cell Res. 34*, 182–212 (1964).

118. W. Engel, W. Krone, and U. Wold, Die Wirkung von Thioguanin, Hydroxylamin und 5-Bromdesoxyuridin auf menschliche Chromosomen *in vitro, Mutation Res. 4*, 353–368 (1967).

119. A. B. Mukherjee, Effect of 5-bromodeoxyuridine on the male meiosis in Chinese hamsters (*Cricetus griseus*), *Mutation Res. 6*, 173–174 (1968).

120. P. C. Nowell, Mitotic inhibition and chromosome damage by mitomycin in human leukocyte cultures, *Exptl. Cell Res.33*, 445–449 (1964).

121. M. M. Cohen and M. W. Shaw, Effects of mitomycin C on human chromosomes, *J. Cell Biol. 23*, 386–395 (1964).

122. M. W. Shaw and M. M. Cohen, Chromosome exchanges in human leukocytes induced by mitomycin C, *Genetics 51*, 181–190 (1965).

123. J. German and J. La Rock, Chromosomal effects of mitomycin, a potential recombinogen in mammalian cell genetics, *Texas. Rep. Biol. Med. 27*, 409–418 (1969).

124. U. H. Ehling, Comparison of radiation – and chemically induced dominant lethal mutations in male mice, *Mutation Res. 11*, 35–44 (1971).

125. M. Buchinger, Cytologische Untersuchungen zur intracellularen Verarbeitung verschiedenen, peroral verabreichter Fluorochrome in Epithelzellen des Magendarmtraktes sowie der Leber und Niere von Ratten, Inaugural Dissertation, Heidelberg (1968).

126. G. Buchinger, Die Wirkung von Trypaflavin allein und in Kombination mit sichtbarem Licht auf die Chromosomen von HeLa-Zellen und menschlichen Leukocyten, *Human Genet. 7*, 323–336 (1969).

127. G. Buchinger, Mutagenicity experiments with mice and human cell cultures: Treatment with an acridine derivative (trypaflavin), *in* "Chemical Mutagenesis

in Mammals and Man" (F. Vogel and G. Röhrborn, eds.) pp. 350–366, Springer-Verlag, Berlin, Heidelberg, New York (1970).

128. N. I. Dishotsky, W. D. Langhman, R. E. Mogar, and W. R. Lipscomb, LSD and genetic damage, *Science 172*, 431–440 (1971).

129. N. E. Skakkeback, J. Philip, and O. J. Rafaelsen, LSD in mice: Abnormalities in meiotic chromosomes, *Science 160*, 1246–1248 (1968).

130. M. M. Cohen and A. B. Murkherjee, Meiotic chromosome damage induced by LSD-25, *Nature 219*, 1072–1074 (1968).

131. M. Hulten, J. Lindsten, L. Lidberg, and H. Ekelund (1968), Studies on meiotic chromosomes in subjects exposed to LSD, *Ann. Génét. 11*, 201–205 (1968).

132. C. F. Somers and T. C. Hsu, Chromosome damage induced by hydroxylamine in mammalian cells, *Proc. Natl. Acad. Sci. 48*, 937–943 (1962).

133. E. Borenfreund, M. Krim, and A. Bendich, Chromosomal aberrations induced by hyponitrite and hydoxylamine derivatives, *J. Natl. Cancer Inst. 32*, 667–680 (1964).

134. D. L. Williams, J. W. Runyan, and A. A. Hagen, Meiotic chromosome alterations produced by progesterone, *Nature 220*, 1145–1147 (1968).

135. W. Sachsse and R. Denk, Chromosomenuntersuchungen aus Lymphocyten und Knochenmark von Psoriasis-Kranken nach Langzeitbehandlung mit Methotrexat, *Arch. Klin. Exptl. Dermatol. 239*, 275–281 (1970).

136. J. F. Jackson, Chromosome aberrations in cultured human leukocytes treated with 8-ethoxycaffeine, *J. Cell Biol. 22*, 291–293 (1964).

137. W. Ostertag, E. Duisberg, and M. Stürmann, The mutagenic activity of caffeine in man, *Mutation Res. 2*, 293–296 (1965).

138. W. Ostertag, Kaffein – und Theophyllinmutagenese bei Zell – und Leukozytenkulturen des Menschen, *Mutation Res. 3*, 249–267 (1966).

139. W. Ostertag and B. J. Greif, Die Erzugung von Chromatidenbrüchen durch Caffein in Leukozystenkulturen des Menschen, *Humangenetik 3*, 282 (1967).

140. W. Ostertag and J. Haake, The mutagenicity in *Drosophila melanogaster* of caffeine and of other compounds which produce chromosome breakage in human cell in culture, *Z. Vererbungsl. 98*, 299–308 (1966).

141. L. I. Waissfeld, The effect of 8-ethoxycaffeine on the cells of embryonic mouse tissues *in vitro*, *Genetika 7*, 103–107 (1967).

142. B. A. Kihlman, G. Odmark, S. Sturelid, and K. Norlen, The relationship between intracellular ATP concentration and the frequency of chromatid aberrations induced by caffeine and 8-ethoxycaffeine in plant root tips and in cell cultures of the Chinese hamster, European Environmental Mutagen Society, Abstracts, First Annual Meeting (May 5–7, 1971), Noordwijkerhout, The Netherlands (1971).

143. M. F. Lyon, J. S. B. Phillips, and A. G. Searle, A Test for mutagenicity of caffeine in mice, *Z. Vererbungsl. 93*, 7–13 (1962).

144. S. Epstein, The failure of caffeine to induce mutagenic effects or to synergize the effects of known mutagens in mice, *in* "Chemical Mutagenesis in Mammals and Man" (F. Vogel and G. Röhrborn, eds.) pp. 404–419, Springer-Verlag, Berlin, Heidelberg, New York (1970).

145. S. S. Epstein, W. Bass, E. Arnold, and Y. Bishop, The failure of caffeine to induce mutagenic effects or to synergize the effects of known mutagens in mice, *Food Cosmet Toxicol. 8*, 381–401 (1970).

146. W. Kulman, H. G. Fromme, E. M. Heege, and W. Ostertag, The mutagenic action of caffeine in higher organisms, *Cancer Res. 28*, 2375–2383 (1968).

147. S. Stone, E. Lamson, Y. S. Chang, and K. W. Pickering, Cytogenetic effects of cyclamates on human cells *in vivo*, *Science 164*, 568–569 (1969).

148. D. R. Stoltz, K. S. Khera, R. Bendall, and S. W. Gunner, Cytogenetic studies with cyclamate and related compounds, *Science 167*, 1501–1502 (1970).
149. M. S. Legator, K. A. Palmer, S. Green, and K. W. Petersen, Cytogenetic studies in rats of cyclohexylamine, a metabolite of cyclamate, *Science 165*, 1139–1140 (1969).
150. S. K. Majumdar and M. Solomon, Cytogenetic studies of calcium cyclamate in *Meriones unguiculatus* (gerbil) *in vivo, Can, J. Genet. Cytol. 13*, 189–194 (1971).
151. J. G. Brewen, F. G. Pearson, K. P. Jones, and H. E. Luippold, Cytogenetic effects of cyclohexylamine and *N*-OH-cyclohexylamine on human leucocytes and Chinese hamster bone marrow, *Nature New Biol. 230*, 15–16 (1971).
152. K. W. Petersen, F. H. J. Figge, and M. S. Legator, Dominant lethal effects of cyclohexylamine in C-57 mice, *Mutation Res.*, in press (1972).
153. M. S. Legator and A. Withrow, Aflatoxin: Effect on mitotic division in cultured embryonic lung cells, *J. Ass. Offic. Agr. Chemists 47*, 1007–1009 (1964).
154. D. A. Dolimpio, C. Jacobson, and M. Legator, Effect of aflatoxin on human leukocytes, *Proc. Soc. Exptl. Biol. Med. 127*, 559–562 (1968).
155. R. F. J. Withers, The action of some lactones and related compounds on human chromosomes, *in* "Proceedings of the Symposium on the Mutational Process, Prague, 1965" (Z. Landas, ed.) Czechoslovakia Academy of Sciences, Prague (1965).

CHAPTER 26

Techniques for Monitoring and Assessing the Significance of Mutagenesis in Human Populations

Howard B. Newcombe

Biology and Health Physics Division
Atomic Energy of Canada Limited
Chalk River, Ontario, Canada

I. INTRODUCTION

Monitoring for mutagenesis in man would not by itself indicate the significance of the changes in mutation rate that are observed. Since the stated purpose of these volumes is to help "avert significant human exposure to mutagenic agents" (p. ix of Volume 1), it is therefore necessary to consider, not only the problem of detecting increases in mutation rate, but also that of assessing their significance in terms of human health. The latter of these two related problems is more difficult than is generally recognized.

The point requires special attention here because it influences substantially the choice of human data which it is appropriate to seek as part of any program of monitoring and protection of man against environmental mutagenesis. It requires emphasis also because so few of the scientists who concern themselves with the mutation process are acquainted with the extent of ignorance about the importance (or lack of importance) of recurrent mutation as a cause of ill health in man. The same might be said of human geneticists interested in the mechanisms of inheritance and disease production. It is as if the tremendous advances which have taken place over the past quarter of a century in knowledge of the mechanisms by which mutations occur and are expressed had competed in some way with interest in the collective importance of mutations as a cause of human ill health.

The question has been asked, "Suppose the mutation rate were doubled, would the effect on human health be serious or would it be trivial?" So far, no satisfactory quantitative answer to this question has been found. And yet the value of gathering information on possible changes in the mutation rate of man will depend almost wholly on having at least an approximate answer to this largely neglected question.

Geneticists commonly turn to current theory to supply such an answer. But relevant theories tend to be poorly substantiated and by no means universally accepted. Thus human data are needed to determine what the mutation-maintained component in human disease is, and any program of monitoring "to avert significant human exposure" would have to be broad enough to include as a major emphasis this sort of activity.

This chapter will consider the kinds of information about the genetics of human populations that are most needed in the above context and some of the ways of obtaining such information. In particular, it will describe methods by which existing computerized population records may be manipulated to yield relevant data in greater quantitites than would be possible by conventional procedures. The essential feature of the approach is that it makes extensive use of existing massive documentation about people by bringing together independently derived records relating to the same persons into individual and family histories of genetically important events, for whole populations.

The future possibilities for more general application of this approach will be considered in relation to protection against mutagenesis and also in the broader context of man's continuing search for an optimum environment.

II. KINDS OF INFORMATION ABOUT HUMAN POPULATIONS NEEDED FOR MONITORING

If the task of monitoring consisted solely of detecting possible changes in the mutation rate in man, one would give thought first to the choice of certain regularly inherited dominant traits for which the mutation rates can most readily be measured, and to the feasibility of using these traits for surveillance. The practical difficulties would undoubtedly be substantial and the sensitivity of detection low. Such an undertaking might perhaps be worthwhile, but the priority is unlikely to be great while major uncertainty exists about the importance of the current mutation rate in terms of the amount of disease caused by it.

While this is in doubt, the greater emphasis must be given to the collection of human data needed to assess the mutational component in genetic disease. In this connection, it is commonly assumed (a) that about six individuals out of every hundred born alive are seriously affected at some time in their lives by known hereditary diseases or congenital disorders of unknown etiology and (b) that perhaps one-sixth to one-half of this "load" of morbidity may be maintained by repeated mutations. However, the uncertainties, especially with respect to the latter assumption, are great.

Most likely to be influenced by an increase in mutation rate is the "monomeric" component of hereditary disease, i.e., those conditions caused by differences in single genes or chromosomes. Information on the magnitude of this component, and the fraction of it maintained by recurrent mutation, is therefore especially needed.

Similar information concerning the less regularly inherited component of genetic disease will also be required, despite current doubts as to whether mutation has any important role in maintaining these conditions. Some geneticists would argue that most of them are multifactorial in origin and that the gene frequencies are determined largely by a balance between opposing selective forces. But the evidence is weak, and data are needed on possible differentials of fertility and mortality in the affected individuals and their relatives.

So far, only one serious attempt has been made to list all genetic diseases occurring in a major geographic area, together with their frequencies among liveborn individuals and in the living population.[1] This list was compiled nearly 15 years ago, however, and has undergone only limited revision since.

At that time, it was usual to regard certain of the less regularly inherited diseases as due to single dominant genes with poor penetrance. This view has since been changing, and there is now reason to believe that a substantial proportion of the illnesses classified in the above reference as due to simple dominant inheritance may be multifactorial in nature. The distinction is important if, as is believed, the multifactorial conditions are relatively insensitive to changes in frequency due to alternations in the mutation rate. (See Section VIB for further discussion of this point.)

The methods of data collection most appropriate for present purposes would therefore be designed to ease the tasks of

    a.   determining disease frequencies without counting multiply ascertained cases more than once,

b. determining the risks of occurrence of genetic diseases in siblings and offspring of affected people, as evidence of monomeric *vs.* multifactorial inheritance,

c. obtaining direct evidence of mutation, e.g., from observations of sporadic cases and from correlations of risks with parental or grandparental ages, and

d. obtaining direct evidence of selection in association with particular genetic diseases, as rendering a major mutational component unlikely for these conditions.

Data of such kinds are normally obtained from individual and family histories of morbidity, mortality, and fertility. Substantial parts of the required information are now being recorded, albeit in fragmentary form, in routine files of which many are readable by computer. Diagnostic information is contained in various health records and in the death registrations, while family relationships, procreative histories, and family histories of mortality are accurately recorded in the vital registration systems.

Since very large files indeed are involved, the extensive use of such records becomes feasible only where the files exist in machine-readable form, or may be readily converted to this form, so as to be manipulated by computers.

III. LIMITATIONS AND ADVANTAGES OF USING EXISTING POPULATION RECORDS

Both the family data from the vital records and the diagnostic data from the health records have inherent merits and limitations. These need to be recognized if appropriate use is to be made of such sources on a large scale, either alone or in conjunction with information obtained in a more conventional *ad hoc* fashion.

Exceedingly accurate and specific information on the compositions of sibships and larger family groupings, for whole vital registration areas, is contained in the accumulated records of marriages, births, and deaths. Such information is readily extractable by computer in a form which will permit the addition of diagnostic data from other sources. The completeness of the family histories obtained in this way is, of course, limited by the movements of people into and out of a registration area, but the limitation is of less importance when information from adjacent areas is consolidated. For many purposes, however, completeness is relatively unimportant, and there is a unique advantage in having family data in the truly enormous quantities represented by whole registration areas or groups of areas.

The diagnostic information contained in the routine health records, such as hospital discharge abstracts and death registrations, is less uniformly reliable and specific than the family information. For the individually rare genetic diseases, such as retinoblastoma and epiloia, the specificity of the diagnoses is often far from adequate, and especially so in their coded forms on the machine-readable versions of these records. Over against this, the collectively more important and readily recognized hereditary conditions, such as cleft palate, congenital dislocated hip, diabetes, and epilepsy, are usually accurately reported, and added assurance may be gained from the histories of repeat diagnoses of the same conditions on different occasions.

Where the diagnostic information on the more readily available routine health records is not adequate for a particular genetic study, the records may still serve as an aid in case finding. In such circumstances, mechanization of the files and linkage of the diagnostic with the family information serve as a major labor-saving device in what would otherwise be a wholly manual study. Thus the mechanized and the manual approaches do not compete with each other but in appropriate circumstances may complement each other.

One is reminded also that the collectively more important part of man's hereditary troubles does not consist of the numerous kinds of individually rare disorders that are inherited in simple Mendelian fashion but is made up instead of the irregularly inherited diseases which tend to be common, are often severe in their manifestations, and are for the most part well reported with specific diagnoses. There is thus a special need to understand the genetic basis of such diseases, and in particular the extents to which mutation and selection are responsible for their continued prevalence.

IV. INFORMATION CONTENT OF THE AVAILABLE RECORDS

The source records of most immediate value are those which are already available routinely in machine-readable form. In Canada, these include (a) the vital records of marriages, births, and deaths, (b) certain specialized disease registers such as those of handicapping conditions and cancer, (c) the discharge summaries from a universal scheme of hospital insurance, and potentially (d) the records from the recently introduced universal medical care insurance. Experience has been gained with linking into individual and family histories all but the last of these four categories of records. [2]

The birth registrations are readily linked with each other and with the parental marriage records to form sibship groupings. This is possible because all of them contain similar family-identifying particulars. In Canada, the machine-readable versions have the surname and maiden surname of the husband and wife, together with their initials and ages, and their provinces and countries of birth in coded form. Some 800,000 such records for the province of British Columbia over the 18-year period 1946–1963 have now been linked into sibship groupings.

Marriage records may also be linked with the prior marriage records of the parents of grooms and brides, and with the marriage records for brothers and sisters of the grooms and brides, since two generations are usually identified on the registration forms. Such linkages, when they span a sufficient period of time, will permit extraction of multigeneration pedigree information from the vital records system. Limited experience with these types of linkage has been gained using magnetic tapes of the British Columbia "marriage index" and "marriage statistical" punchcards which are produced routinely, together with specially prepared punchcards containing the names of the parents of grooms and brides.

With such linkages, entries of information from a given marriage record will appear in three locations in the linked file. In one of these locations, the entry represents an event to a male member of the sibship group into which the groom was born, in another an event to a female member of the sibship into which the bride was born, and in the third an event to the prospective parents of a potential new sibship group.

In the British Columbia study, the sibship groupings of records are sequenced in order of the paternal surname followed by the maternal maiden surname, both in phonetically coded form. Marriage entries in each of the three locations in this file are cross-referenced to each other by means of the three pairs of surname codes representing the respective locations in the file sequence, together with the marriage registration number which is common to all three entries. In this manner, it is possible to store, in a single linear array, multigeneration pedigree information of unlimited complexity derived from the marriage registrations.

Diagnostic information from the death registrations, special registers of handicapping conditions, and hospital insurance discharge summaries is incorporated into the sibship groupings of vital records by linking these first with the appropriate birth registrations. For the purpose of carrying out this linkage, reliance is placed mainly on names and birthdates. Birthplace would be equally important if available but is frequently lacking from hospital discharge records and handicap records. However, in the case of children, information on next of kin or the person in whose name the family is enrolled for hospital insurance may serve to resolve ambiguities.

Diagnostic information, for children born in British Columbia over the 6-year period 1953–1958, has now been entered into the sibship groupings of birth records linking with these the records of deaths and registered handicaps. Limited experience has been gained also with linkage of hospital insurance discharge summaries, for children, with the corresponding birth records. Since the routinely produced machine-readable versions of the discharge summaries lack names, these and other identifying particulars were especially keypunched for the study and use was made of the hospital code number and patient admission number to bring the identifying and statistical information together.

No attempt has been made so far to link into sibship groupings the diagnostic information from the medical care insurance records of visits to doctors' offices. This information does exist in Canada, in machine-readable form, for virtually all such visits. Furthermore, it is grouped routinely for parents and dependent offspring under the account numbers of the persons who pay the family premiums.

The advantages of compiling cumulative health histories from routine records have been widely discussed in terms of a variety of administrative and statistical uses, for studies of the efficiency of health care delivery and the efficacy of treatment methods, and for research in epidemiology[3-10]. A more extensive bibliography is contained in Wagner and Newcombe. [11]

V. METHODS OF RECORD LINKAGE

Large-scale compilation of personal and family health histories could be greatly simplified if use were made on all vital and health records of some universal system of personal identity numbers, similar to the Social Security Number in the United States or the Social Insurance Number in Canada. There is still great resistance on the North American continent, however, to any rapid and major extension of the purposes for which these numbers may be employed. For the future, public attitudes in such matters may be expected to evolve along with appropriate safeguards to personal privacy and increased recognition of specific advantages to be gained.

In the meantime, most linkages of routine records of the kinds required for studies of human population genetics will have to be based primarily on the names of the people concerned, their birthdates, birthplaces, and so on. All of these items of identifying information are subject to errors and to variations in the manner in which they are entered on successive documents pertaining to the same person. Despite this unreliability of the individual items, however, collectively they may possess very substantial discriminating power.

The procedures for linkage of records by computer are designed to make nearly maximum use of the discriminating power inherent in the available identifying information. The description that follows will be nontechnical, but the reader is referred to a number of articles which discuss more fully the theory and practice of "nominal" record linkage. [12-20]

Linkage of independently derived records relating to the same person or family involves two steps, a searching step and a matching step. In the searching step, the aim is to minimize failures to bring potentially linkable records together for comparison, whereas in the matching step both false linkages and failures to link must be guarded against.

In the searching step, where large numbers of records are to be linked, the degree of success will depend on the choice of the most reliable items of identifying information for sequencing the files. To limit the cost, these should contain sufficient discriminating power to divide the file finely enough so that each incoming record will not have to be compared with an excessive number of records already in the file. For linkage of the British Columbia marriage and birth records into sibship groups, we made use of a sequence based on a phonetic code (the Russell Soundex code) of the husband's surname followed by that of the wife's maiden surname. In practice, the coding was found to avoid two-thirds of the failures to bring potentially linkable records together for comparison which would have occurred, as a result of variant spellings, if a strictly alphabetic sequence had been used. In a file of over 100,000 marriages, the subdivision by this double surname code is fine enough so that on the average an incoming birth record has to be compared with only 1.6 marriages.

In the matching step, the computer must exercise something akin to human judgment when comparing items of identifying information, some of which agree and some of which disagree. Initially, this was regarded as a major hurdle until it was realized that the rules of human judgment in such situations are not too complicated to be followed by a computer. Furthermore, the information needed by the computer in order to assess the odds in favor of a genuine match, or against it, is not too bulky to be stored in the form of look-up tables as part of the linkage program.

In essence, the computer needs information on the discriminating powers of the various agreements and disagreements, and given this it can compute the required odds. Uncommon names, initials, and birthplaces will obviously carry greater discriminating power than their more common counterparts. Similarly, discrepancies that occur only rarely will argue more strongly against a genuine linkage than will the more frequent kinds of discrepancy. For details of the mathematical treatment, see Newcombe et al.,[12] Newcombe and Kennedy,[13] Newcombe,[14] and Fellegi and Sunter.[20]

In practice, we have used look-up tables of the positive weights to be attached to different given names, different letters of the alphabet when used as initials, different provinces and countries of birth, and so on, together with look-up tables of the negative weights to be attached to the various kinds of disagreements. When a pair of records is compared, the positive and negative weights are summed, and, since the individual weights are generally uncorrelated with each other, this sum serves to indicate in quantitative fashion the likelihood of a linkage or a nonlinkage. The whole process of comparison and assessment can be carried out very rapidly by computer.

The approach sounds complicated but is in fact exceedingly simple and flexible in its application. For example, agreement of the rare initials Z or Q, for a male, will obviously carry much greater weight than agreement of the more common initials J or R. Similarly, disagreement with respect to the recorded sex, which is hardly ever wrongly stated, will virtually rule out the possibility that the two records relate to the same person, whereas disagreement with respect to occupation or the place of the two events will be common and may indicate merely a change of employment or place of residence. All of this can be quantitated readily using information from the files themselves.

Computer programs to carry out the linkages in this manner have been extensively used in connection with the following: in the British Columbia records; in the Oxford Hospital Region in Britain to prepare personal health histories which include birth, repeated events of hospitalization, and death; and to lesser extents in Maryland, Pavia, Hawaii, and Iceland for a variety of purposes. As applied to the vital records, the accuracies and speeds are high and the cost of the linkage proper is exceedingly low once the files have been prepared in an appropriate format and sequence.

VI. RELEVANT DATA OBTAINED BY RECORD LINKAGE

Examples will be given of the uses of record linkage in attempts to (a) estimate the combined importance of the known genetic diseases and of the congenital and other diseases that may be hereditary or partially so, (b) distinguish the monomeric from the multifactorial parts of this total load of disease, and (c) identify a mutation-maintained and a selection-maintained component in these two parts.

The examples will serve in part to illustrate certain difficulties in thus neatly partitioning our load of hereditary troubles and the needs for data that have not as yet been obtained by any method. They will show, moreover, that whereas conventional approaches may be desirable to achieve the diagnostic specificity required with respect to some of the rarer genetic diseases, the total volume of diagnostic, reproductive, and family data required in the context of a monitoring program is unlikely ever to be achieved by conventional methods alone, i.e., without efficient and extensive use being made of existing routine sources.

A. Amount of Hereditary Disease

Without adequate knowledge of the numbers of cases of hereditary diseases that occur in the population, it would be difficult indeed to estimate the likely consequences of a given increase in the mutation rate. And to actually detect an increase in mutation rate in man, knowledge of the numbers and kinds of such diseases will have to be exceedingly precise and complete. However, despite widespread interest in the diverse forms of hereditary disease, only limited interest has been shown in determining their frequencies and measuring their combined impact on human health.

No single file of records and no single *ad hoc* survey are likely to uncover all of the cases of any major category of disease in a population. It is not unusual therefore to search manually all relevant record files when attempting to identify cases of any important disease, such as blindness or deafness, in a given area. When such files are already in machine-readable form, or can readily be made so, computers may be regarded as capable of performing what would otherwise be the clerical tasks of collecting together the data from the diverse sources and ensuring that the same affected individual is not counted more than once.

The approach has been used to determine the numbers of children in Nova Scotia suffering from hereditary and congenital defects, as represented by the records of hospitalization and deaths in the first year of life, and their appearance in a provincial register of handicapped children. About 48 children per 1000 liveborn were found by this means to be affected by conditions which may be regarded as hereditary, congenital, or both, and which are detectable early in life. [21–22]

Since rare hereditary diseases are commonly included together with other disorders in the diagnostic categories recognized by the international classification of diseases and causes of death most often used in the machine records, [23] it is recognized that manual access to the original paper documents will be needed for some purposes. Thus a combination of computer methods with manual procedures will be needed to make best use of the information available at this time. For the future, however, diagnostic detail may be expected to become increasingly available from computerized record-keeping systems used by hospitals.

B. Distinguishing the Monomeric and Multifactorial Parts

Of the diseases caused by single gene differences, it is generally agreed that the regularly inherited dominants with severe effects are probably maintained by recurrent mutations. Whether this is true of recessive diseases is less certain, and it is quite unlikely to be true of the multifactorial diseases. The distinction between the monomeric and multifactorial traits is thus important for assessment of the significance of any increase in the mutation rate. This distinction is most readily based on the patterns of inheritance, and in particular on the incidence among close relatives of affected people. Fortunately, the risks to siblings of affected people of virtually all recorded diseases are obtainable in quantity from the family-linked vital and health records, and other sorts of data are potentially available.

Theoretical predictions of the risks to siblings of affected people, for what may be a major part of the multifactorial diseases of man, have been derived by Edwards [24] and by Falconer [25] on the basis of a "quasi-continuous" or "threshold" model. These predictions have been subjected to an extensive empirical test by Trimble [26] based on a large number of sibship groupings of birth weights, derived by record linkage from the British Columbia vital records. Use was made of the information on the "risks" of exceeding various arbitrarily chosen "thresholds" for birth weight among siblings of "affected" infants. Reasonable agreement has been found between the observed and expected "risks" of repeat "occurrences" in the same sibships.

The family-linked British Columbia files have in addition yielded sib risk data for such conditions as strabismus, clefts of the palate and lip, clubfoot, heart malformations, spina bifida, and epilepsy of children. [27] It would thus be technically feasible to compare systematically for a wide range of diseases the actual risks among siblings of affected people with those expected for multifactorial quasicontinuous inheritance.

The need for this kind of information becomes apparent when attempts are made to determine the combined frequency of the regularly inherited dominant diseases, which are of special interest as the only group of conditions likely to increase in direct proportion to an increase in the mutation rate. On the basis of a list of such diseases prepared for the United Nations Scientific Committee on the Effects of Atomic Radiation [1] (see pp. 197–200), their combined frequency has sometimes been taken as affecting seriously approximately 1 % of liveborn individuals at some time in their lives. However, at the time when this list was prepared dominant inheritance of disease with single-gene causation was regarded as more common than it is now believed to be. The true figure may thus be substantially lower than the 1 %, perhaps by as much as tenfold.

Current data are unfortunately inadequate to resolve the uncertainty. The point may be illustrated with the condition hydrocephalus, which is included in the list of dominants and accounts for one-fifth of the total cases in this category. From the British Columbia data, the risk to a sibling of an affected person is less than 2%, so it clearly cannot be inherited as a simple dominant (or even as a simple recessive) in any substantial proportion of cases. In fact, similar uncertainty exists for many of the more common conditions that make up the bulk of the list (Table 1).

The difficulties of compiling such a list of dominant diseases which includes their frequencies in a birth cohort become particularly apparent when attempts are made to derive appropriate data from the literature. For diseases that are sometimes inherited in a simple dominant

TABLE 1.   List of Autosomal Dominant Traits, in Descending Order of Frequency, Based on the Population of Northern Ireland [a]

	Disease	Frequency per 10^6 live births	Footnotes
1.	Cataracts, senile and presenile	2000	b
2.	Hydrocephaly, internal obstructive	1230	c
3.	Alopecia areata	700	d
4.	Nystagmus, familial idiopathic	700	e
5.	Cystic disease of lungs	500	f
6.	Choroidal sclerosis	500	g
7.	Multiple exostoses	400	h
8.	Neurofibromatosis	300	i
9.	Colobomata	250	j
10.	Ataxia, dominant hereditary, including Friedrich's	200	k
11.	Porphyria, dominant detectable	200	l
12.	Cataracts, congenital	160	m
13.	Fundal dystrophies	150	
14.	Retinitis pigmentosa	150	
15.	Corneal dystrophies	140	
16.	Spastic diplegia	100	
17.	Polyposis of colon, multiple	100	
18.	Epidermolysis bullosa	100	
19.	Teglangiectasis hemorrhagica	100	
20.	Megacolon	100	
21.	Glaucoma, infantile and juvenile	100	
22.	Hypermetropia	100	
23.	Muscular dystrophies, dominant	100	
24.	Urticaria pigmentosum	90	
25.	Arachnodactyly	60	
26.	Osteogenesis imperfecta	60	
27.	Spherocytosis, alcholuric jaundice	60	
28.	Thrombocytopenia, chronic recurrent	60	

TABLE 1.   cont'd

29.	Aniridia	60
30.	Retinoblastoma	58
31.	Deaf-mutism, deafness totally hereditary	46
32.	Polycythemia vera	45
33.	Peroneal muscular atrophy	40
34.	Dystrophia myotonica	40
35.	Xanthoma tuberosum multiplex	40
36.	Diabetes insipidus	40
37.	Tylosis palmaris et plantaris	35
38.	Anhidrotic syndrome	34
39.	Ectrodactyly, including all "split hand"	30
40.	Osteitis deformans	30
41.	Craniofacial dysostoses	30
42.	Epiloia, tuberose sclerosis	30
43.	Cephalofacial hemangiomatosis	30
44.	Achondroplasia	28
45.	Muscular dystrophy, limb girdle	25
46.	Willebrand's disease	25
47.	Hypertelorism	20
48.	Myositis ossificans	20
49.	Pityriasis rubra pilaris	20
50.	Keratoconus	20
51.	Deafness, perception, early dominant	12
52.	Deafness, absence or atresia of external auditory meatus	12
53.	Huntington's chorea	10
54.	Optic atrophy	7
55.	Brachydactyly, major	6
56.	Deafness and cataract, severe early	6
57.	Subluxation of lens, primary	6
	Total	9555

[a] Adapted from the 1958 "Report of the United Nations Scientific Committee on the Effects of Atomic Radiation" [1] Diseases 1–10 account for 75% of the affected individuals represented in the table as a whole. It is apparent that the usually accepted figure of 1% of liveborn individuals severely affected by simple dominant diseases may greatly overestimate the true impact of dominant mutation. To resolve the uncertainties, it will be more important in the future to obtain systematic family risk data for recognizable categories of disease, of known frequencies as represented in such a table, than to seek the patterns of inheritance for a finer diagnostic breakdown where the frequencies of conditions are not known.

[b] Although sometimes described as apparent dominant with poor penetrance, cataracts occur also in association with other diseases such as diabetes. What proportion of the senile and presenile sorts represents regularly inherited dominant conditions appears uncertain.

[c] British Columbia data indicate a risk among siblings of affected people in the vicinity of 3% ruling out simple inheritance except perhaps in rare cases.

[d] Although autosomal dominant inheritance has been postulated, the risk among close relatives of affected people is about 20%, indicating either poor penetrance, multifactorial inheritance, or an admixture of nongenetic causes.

[e] Reported to have many causes, not all hereditary.

[f] Mode of inheritance not well established.

[g] Mode of inheritance not well established, but dominant causation regarded as less frequent than recessive.

[h] Only a minority troublesome.

[i] About 40% show dominant inheritance, and not all cases are seriously affected.

[j] Vary from slight defect of iris to major defect extending beyond the iris; irregular dominant inheritance with 20–30% of descendants affected.

[k] A mixed group which is now thought to exclude Friedrich's ataxia; modes of inheritance not all well established.

[l] A mixed group; modes of inheritance not all well established.

[m] A mixed group; often recessively inherited, sometimes not hereditary.

fashion, there is a tendency to report mainly on those families which show regular inheritance, without providing data on the frequency with which the disease is not so inherited (e.g., see McKusick [28]). There is a deficiency of studies in which the frequencies of broad groups of genetic diseases, and the risks among close relatives of affected people, have been studied simultaneously in whole populations of substantial size. Until such information can be obtained systematically, it may continue to be difficult to arrive at an improved estimate of the amount of disease due to the regularly inherited dominants.

C. Tests for a Mutation-Maintained Component

Direct evidence of a mutationally maintained component in human hereditary disease may be sought in at least two ways: (a) from pedigree studies designed to identify the sporadic cases of known dominant conditions and (b) from correlations of the risks of disease with advancing parental (or grandparental) age at the time of conception or birth. The studies of sporadic cases are readily interpretable only where the disease is regularly expressed when the causal gene is present. And the parental age correlations provide convincing evidence of a pile-up of mutations with advancing age only where the contribution of environmental variables to the correlation can be excluded.

For the purpose of pedigree studies to identify sporadic cases of dominant diseases, the routine records in their linked form would provide information on family composition for whole populations and would serve as an aid in finding the cases. While the extent of the labor saved would be considerable, it is probable that substantial further work of conventional kinds would be needed where the coded diagnoses lacked the required specificity. This would be true also where it was decided to extend the family histories to include events of ill health not represented in the routine records for the region, e.g., because of the movements of people into and out of the area under study.

For the purpose of investigating correlations of risks of genetic diseases with advancing parental age, the advantages of working through the linked vital and health records are enormous. It becomes possible to make use of the parental ages as stated on the birth records for a whole registration area and to identify enough of the affected individuals in the area so that they cannot be regarded as drawn from an unrepresentative subgroup of the population. One major source of bias is thus avoided.

Experience has been gained with such studies using the British Columbia records. An effect of advancing maternal age has in this manner been demonstrated for mongolism, cerebral palsy, and congenital malformations of the circulatory system. [29,30] Since the birth order of a child is contained in its birth registration, it was possible to remove the effects of birth-order differences when comparing different maternal ages. It is not always apparent, of course, whether a given maternal age effect is mutational in origin, and for diseases in which there is a substantial environmental component the influence of an aging maternal physiology may be the true cause.

This particular difficulty of interpretation does not arise where an effect of advancing paternal age is sought. Positive correlations with father's age independent of mother's age have been found, using the linked family files of British Columbia records, for congenital malformations as a class and for a group of conditions known in the international classification as "other congenital malformations of the nervous system and sense organs." [31] This latter group includes congenital cataract, various eye anomalies, and maldevelopment of the brain; it excludes anencephalus, meningocele, and hydrocephalus. The observed association could perhaps indicate that the likelihood of mutation in the paternal sperm increases with advancing age, but the evidence is not conclusive, for reasons which will become apparent.

To illustrate the difficulty of interpreting the paternal age data where a condition is not known to be determined wholly by genetic factors, it should be noted that infant and child deaths from pneumonia and bronchitis were also correlated with paternal age in the above study. It is unlikely indeed that this association is genetic in origin. The true explanation became apparent only when it was found that there was an unusually high number of older men among the North American Indians who had married younger women and had children by them and that the children had the high risk of death from respiratory disease characteristic for Indians. The correlation disappears in data that are homogeneous with respect to Indian or non-Indian origins.

There appear to be only two ways of avoiding or minimizing biases of this sort. One is to restrict such studies to those hereditary diseases which are uninfluenced in their expression by environmental factors, and the other is to control for all social variables which may be correlated with environmental differences that affect the expression of the diseases. With either approach, very substantial populations must be investigated. In the first approach, this is because only a few rare dominant diseases qualify for study. In the second, where the tabulations must be broken down into subsets homogeneous for a multiplicity of social variables, large quantities of data are needed to permit comparisons within subsets.

Thus if indications of an important mutational component in human genetic disease are to be obtained from studies of sporadic cases or from correlations of risks with parental (or grandparental) ages, data from large populations will be required. Only by making substantial use of linked vital and health records is the appropriate scale of investigation likely to be achieved at reasonable cost.

D. Tests for a Selection-Maintained Component

For a proportion of the single-gene traits, and probably for most of the multifactorial traits, selection is either wholly or partially responsible for the continued prevalence of the causal genes. Where such a selective effect can be demonstrated, it becomes much less likely that an increase in mutation rate would greatly affect the incidence of the disease. Refinement in the detection of relatively small differentials of fertility and mortality depends mainly on the large-scale availability of reproductive histories such as are yielded by linkage of the vital records. Experience in the study of selection with respect to both monomeric and multifactorial traits has been gained using this approach.

For example, according to current theory the continued prevalence of the Rh-negative gene in Western populations may be due to reproductive compensation or overcompensation on the part of mothers whose children have died of erythroblastosis. A direct test of the theory was not carried out, however, until the reproductive histories of the mothers of erythroblastotic children could be obtained prospectively by linking the British Columbia birth records into sibship groups. Because the linkage involved all births in the population under study, i.e., about 200,000 over a 5-year period in a population of nearly 2 million people, an unusual degree of refinement was possible.[32,33] The controls in the study consisted of all the mothers of children born in the same period as the erythroblastotic children, and the comparisons of the maternal fertilities could be made specific for the ages and parities of the mothers. This test, incidentally, did not indicate any reproductive compensation for erythroblastosis.

Selection, of course, could be acting in a different manner and could involve almost any stage in the life cycle. With virtually complete family histories of procreation and mortality available for whole populations in machine-accessible form, such tests for selective differentials may be applied in various ways with a degree of precision far beyond what would ever be possible for studies of more conventional design.

Similar tests for selection in relation to the multifactorial conditions are likewise possible on a large scale, although the results are more difficult to interpret. In the British Columbia study, for example, a higher than expected reproductive rate was found following deaths of infants from postnatal asphyxia and atelectasis (i.e., failure of the lungs to expand at birth). No such effect was observed in mothers of infants dying of hemorrhagic disease. Furthermore, among mothers of stillborn children the subsequent fertility was considerably below that of the control mothers except for a brief period of overcompensation during the first year.

To interpret such correlations, additional information may be needed. In the case of postnatal asphyxia, for example, further data are required to distinguish reproductive compensation following the deaths from a simple tendency for the disease to occur in population subgroups that are characteristically more fertile than average.

Despite such complexities of interpretation, better insight into the role of selection in maintaining the prevalence of multifactorial diseases is sorely needed in the present context to indicate those for which mutation is unlikely to be a major cause. The ultimate in refinement will only be reached with an abundance of pedigree information, sufficient to permit comparisons within data subsets homogeneous for a multiplicity of social variables. To obtain this from the linked vital and health records is not too difficult, whereas *ad hoc* studies of conventional design will probably always be inadequate for the purpose.

VII. FUTURE ACCESSIBILITY OF DIAGNOSTIC DATA

Record linkage should not be thought of merely as a genetic tool. For the approach to be developed more fully in the future, a multiplicity of uses of the linked files needs to be recognized, extending far beyond the special interests of the geneticists. The maintenance of large files of personal health records in a fashion that renders linkage potentially possible is now coming to be regarded as simply good housekeeping, and the uses of linkage for a wide variety of statistical and administrative purposes are being increasingly discussed. In particular, the many potential uses of lifetime health histories derived from existing computerized records are attracting special attention.

One inference that might be drawn from this for the geneticists is that a current involvement on their part during the course of such developments may influence substantially future accessibility of human genetic data. This will be especially important if geneticists are to carry responsibility for monitoring and assessing possible adverse genetic trends, including those due to environmental mutagenesis. The information content of the routine records will undoubtedly become more readily accessible in the future, but the patterns of access will not necessarily be optimal for such purposes unless geneticists themselves participate in the development of these patterns.

VIII. MUTAGENESIS AND THE SEARCH FOR AN OPTIMUM ENVIRONMENT FOR MAN

Mutagenesis is one of a number of harmful environmental effects against which society is currently seeking to protect itself, and the methods of assessment and monitoring may well be similar for many of these. Efficiency in such activities could thus depend on an appropriate pooling of effort and the joint use of existing data resources including the source files of personal health records and of family reproductive histories. The point will be made that both the diagnostic and the family data are needed for more than one type of environmental assessment and monitoring.

The thalidomide incident has alerted people to the possible teratogenic effects of drugs and other chemical constituents of our changing environment. Carcinogenic agents encountered in both our working and living environments, and unwittingly or unavoidably associated with medical diagnostic and therapeutic measures, are a continuing cause for concern. Geneticists, moreover, have in the past been as concerned about possible adverse effects of selection pressures, where these are modified by circumstances prevailing in advanced societies, as they now are about mutagenesis.

For these various hazards, assessment and monitoring involve the use of diagnostic data and the following up of individuals and families to determine what happens to them. Registries of congenital anomalies, set up to monitor for possible teratogens in the environment, and cancer registries both have much in common with genetic registries (e.g., see "Registry,"[34] Banister,[35] and Klemetti and Saxén[36]), a point which has been made earlier by Crow.[37] For the future, these three sorts of data resource should all make increasing use of information which is recorded routinely and is available in machine-readable form.

Family data are important in all three connections. The effects of teratogens will be distinguished from those of genetic origin, mainly on the basis of family data. Carcinogenic effects may need to be looked for among members of the families of individuals who are either already affected or have been exposed to special risks (e.g., in spouses of persons with cancer of the prostate or cervix, and in children and husbands or wives of heavy smokers). And family reproductive data are as important to geneticists concerned about unfavorable selection pressures in certain environments as to those interested in mutagenesis.

The needs for refinement in the methods of assessment and monitoring are substantially similar as relating to all of the above problems. Thus if efficiency in these matters is to be achieved, it will be through an integrated approach on the part of geneticists and epidemiologists, who will share organized data sources and a common interest in achieving an optimum environment for man.

IX. CONCLUSIONS FOR GENETICISTS

The complexities that have been described pose a special problem for the "pure" geneticist who wants to protect the quality of the human germ plasm but cannot yet say how important it is to do so. I should like to indicate briefly some of the options open to him.

If he is convinced that the matter is important enough to merit serious personal involvement on his part, with the directly relevant work of assessing the impact of mutations on man and looking for possible changes in the mutation rate, he will be led far from the basic research on the mechanisms of mutation in which he was first interested. If, however, he leaves the task to others, there is a reasonable likelihood that it may not get tackled at all. Social concern in such matters has a tendency to arise first among basic scientists as an outcome of their own research, and it is often not shared in full measure by their more "applied" colleagues.

Twenty-five years ago it was reasonable to hope that, with adequate emphasis on basic studies of the process of induced mutation, answers to practical questions about the levels of protection against ionizing radiation needed by man for genetic reasons would be obtained as a by-product. Despite the scientific advances that have been made, hindsight now gives less reason for optimism of this particular kind. We are still exceedingly unsure how much disease a given increase in the mutation rate would cause, and some of the more empirical but strictly relevant approaches to this problem have in the meantime been neglected.[38] If the task of protecting man's germ plasm is important, as many geneticists including the author intuitively feel, a shift in emphasis to some of the more strictly relevant lines of research would seem appropriate at this point in time.

X. REFERENCES

1. "Report of the United Nations Scientific Committee on the Effects of Atomic Radiation," pp. 197–200, General Assembly Official Records: Thirteenth Session, Supplement No. 17 (A/3838), New York (1958).
2. H. B. Newcombe, Present state and long term objectives of the British Columbia population study, in "Proceedings of the Third International Congress of Human Genetics," pp. 291–313, Johns Hopkins Press, Baltimore (1967).
3. E. D. Acheson, "Medical Record Linkage," Oxford University Press, London (1967).

4. E. D. Acheson (ed.), "Record Linkage in Medicine," E. and S. Livingstone Ltd., Edinburgh (1968).

5. "Health Research Uses of Record Linkage in Canada," Report No. 3 of the Medical Research Council of Canada, Ottawa (1968).

6. "Report on the Activities of the Ontario Council of Health, June 1966 to December 1969," Ontario Department of Health, Toronto (1969).

7. "Report of the Ontario Council of Health on Health Statistics," Part 1, Annex G, Ontario Department of Health, Toronto (1969).

8. "Report of the Ontario Council of Health on Health Care Delivery Systems. Role of Computers in the Health Field. Supplement No. 9," Ontario Department of Health, Toronto (1970).

9. "Use of Vital and Health Records in Epidemiological Research," National Center for Health Statistics, Public Health Service Publication No. 1000, Series 4, No. 7, Government Printing Office, Washington (1968).

10. "Epidemiological Methods in the Study of Chronic Diseases," World Health Organization Technical Report Series No. 365, World Health Organization, Geneva (1967).

11. G. Wagner and H. B. Newcombe, Record linkage, its methodology and application in medical data processing, *Meth. Inform. Med. 9*, 121–138 (1970).

12. H. B. Newcombe, J. M. Kennedy, S. J. Axford, and A. P. James, Automatic linkage of vital records, *Science 130*, 954–959 (1959).

13. H. B. Newcombe and J. M. Kennedy, Record linkage: Making maximum use of the discriminating power of identifying information, *Commun. Ass. Computing Machinery 5*, 563–566 (1962).

14. J. M. Kennedy, H. B. Newcombe, E. A. Okazaki, and M. E. Smith, "Computer Methods for Family Linkage of Vital and Health Records," Report No. AECL-2222, p. 30, Atomic Energy of Canada Ltd., Chalk River, Ontario (April 1965).

15. H. B. Newcombe, Record linking: The design of efficient systems for linking records into individual and family histories, *Am. J. Human Genet. 19*, 335–359 (1967).

16. J. M. Kennedy, File structure for automatic manipulation of linked records, *in* "Record Linkage in Medicine" (E. D. Acheson, ed.) pp. 109–117, E. and S. Livingstone Ltd., Edinburgh (1968).

17. D. Nitzberg, Results of research into the methodology of record linkage, *in* "Record Linkage in Medicine" (E. D. Acheson, ed.) pp. 187–201, E. and S. Livingstone Ltd. Edinburgh (1968).

18. W. Phillips, Record linkage for a chronic disease register, *in* "Record Linkage in Medicine" (E. D. Acheson, ed.) pp. 120–151, E. and S. Livingstone Ltd., Edinburgh (1968).

19. M. R. Hubbard, "Computer Systems for Medical Record Linkage," Thesis submitted for the degree of Bachelor of Science, p. 157, University of Oxford, Magdalen College (1969).

20. I. P. Fellegi and A. B. Sunter, A theory for record linkage, *J. Am. Stat. Ass. 64*, 1183–1210 (1969).

21. H. B. Newcombe, Value of Canadian hospital insurance records in detecting increases in congenital anomalies, *Can. Med. Ass. J. 101*, 121–128 (1969).

22. H. B. Newcombe, Pooled records from multiple sources for monitoring congenital anomalies, *Brit. J. Prevent. Soc. Med. 23*, 226–232 (1969).

23. "Eighth Revision International Classification of Diseases, ICDA Volumes 1 and 2, Adapted for Use in the United States," Public Health Service Publication No. 1693, Government Printing Office, Washington (1967).

24. J. H. Edwards, The simulation of mendelism, *Acta Genet. 10*, 63–79 (1960).

25.   D. S. Falconer, The inheritance of liability to certain diseases, estimated from the incidence among relatives, *Ann. Human Genet.* 29, 51–71 (1965).

26.   B. K. Trimble, An empirical simulation of quasi-continuous inheritance using human birthweight data, Thesis submitted for the degree of Master of Science, pp. 46+24 McGill University, Montreal (1971).

27.   H. B. Newcombe, Familial tendencies in diseases of children, *Brit. J. Prevent. Soc. Med.* 20, 49–57 (1966).

28.   V. McKusick, "Mendelian Inheritance in Man," 3rd ed., Johns Hopkins Press, Baltimore (1971).

29.   H. B. Newcombe and O. G. Tavendale, Maternal age and birth order correlations. Problems of distinguishing mutational from environmental components, *Mutation Res.* 1, 446–467 (1964).

30.   H. B. Newcombe, Screening for effects of maternal age and birth order in a register of handicapped children, *Ann. Human Genet.* 27, 367–382(1964).

31.   H. B. Newcombe and O. G. Tavendale, Effects of father's age on the risk of child handicap or death, *Am. J. Human Genet.* 17, 163–178 (1965).

32.   H. B. Newcombe and P. O. W. Rhynas, Child spacing following stillbirth and infant death, *Eugen. Quart.* 9, 25–35 (1962).

33.   H. B. Newcombe, The study of mutation and selection in human populations, *Eugen. Rev.* 57, 109–125 (1965).

34.   Genetically determined disease in the Registry case load, Section III, *in* "Registry for Handicapped Children and Adults, Annual Report 1969," Special Report 118, pp. 29–32, Division of Vital Statistics, Department of Health Services and Hospital Insurance, Province of British Columbia, Victoria (1970).

35.   P. Banister, Congenital malformations: Preliminary report of an investigation of reduction deformities of the limbs, triggered by a pilot surveillance system, *Can. Med. Ass. J.* 103, 466–472 (1970).

36.   A. Klemetti and L. Saxén, "The Finnish Register of Congenital Malformations, Organization and Six Years of Experience," Report No. 9, p. 30, Health Services Research of the National Board of Health in Finland, Helsinki (1970).

37.   J. F. Crow, Human population monitoring, *in* "Chemical Mutagens" (A. Hollaender, ed.) Vol. 2, pp. 591–605, Plenum Press, New York (1971).

38.   H. B. Newcombe, Effects of radiation on human populations, *in* "Human Genetics. Proceedings of the Fourth International Congress of Human Genetics, Paris, September 1971," Excerpta Medica Foundation, Amsterdam, pp. 45–57.

CHAPTER 27

Specific-Locus Mutational Assay Systems for Mouse Lymphoma Cells

Donald Clive, W. Gary Flamm, and James B. Patterson

Mutagenesis Branch
National Institute of Environmental Health Sciences
National Institutes of Health
Research Triangle Park, North Carolina

I. INTRODUCTION

For several years, it was generally believed that tissue culture cells derived from mammalian sources could not be mutagenized by exposure to known mutagens (Szybalski *et al.*, 1964; Morrow, 1964). However, in 1968, independent studies by Chu and Malling (1968*a*, *b*) and Kao and Puck (1968) revealed that gene mutations are induced when Chinese hamster cells are incubated for short periods in the presence of alkylating agents. Additional studies proved that the spontaneous mutation rate for mammalian tissue culture cells is actually several orders of magnitude lower than that assumed in the 1964 reports. Based on this knowledge, it became easy to understand how Szybalski and Morrow failed to demonstrate an enhancement of mutant frequency upon treatment of cells with mutagens; i.e., the background mutant frequency was too high to observe an increase arising from the induction of new mutants.

During recent years, extensive studies of the HGPRT (hypoxanthine-guanine phosphoribosyl transferase) locus in humans (Lesch and Nyhan, 1964; Seegmiller *et al.*, 1967) and in mammalian cells grown in culture (Lieberman and Ove, 1960; Davidson *et al.*, 1962; Szybalski and Szybalska, 1962; Littlefield, 1963; Morrow, 1964, 1970; Chu and Malling, 1968*a*, *b*;

Kao and Puck, 1968) have attributed "classic" status to this X-linked gene. Indeed, it has become a standard against which all other development in this field must be measured. For this reason, we felt it necessary to first acquaint ourselves with the HGPRT locus before striving to develop other loci in the Fischer L5178Y mouse lymphoma-cell system (Fischer, 1958).

Initially, we became attracted to the potential of L5178Y. cells because of a number of characteristics which would seem to favor their use in an *in vitro* mammalian cell mutational assay system. These include growth in suspension culture, a short generation time of 11 hr, and a cloning efficiency of up to 100% in a simple soft-agar cloning medium.

We describe in this chapter our experience with the X-linked HGPRT locus and compare some of the results derived from studies with this locus to those obtained with the newly developed thymidine kinase locus, which we believe to be autosomal. Details of the methodology are described, and an attempt is made to compare the utility of L5178Y cells to that of other cell lines used for specific-locus mutational assays.

II. HGPRT LOCUS

The hypoxanthine-guanine phosphoribosyl transferase (HGPRT) locus in mammalian cells lends itself exceedingly well to the detection and quantitation of forward mutational events because the enzyme specified at this locus (i.e., the phosphoribosyl transferase) is solely responsible for the sensitivity of a cell to certain drugs, specifically to 8-azaguanine (azg), 6-mercaptopurine (6-MP), and certain other purine analogues of guanine or hypoxanthine. The reason for this sensitivity has been thoroughly reviewed by Chu (1971a), but briefly stated it is the fact that the drugs listed above are essentially innocuous unless enzymically converted through ribose phosphorylation (by HGPRT activity) to nucleotides, which can then inhibit purine synthesis or be incorporated into nucleic acids. These latter events are lethal to the cells in which they have occurred. Thus HGPRT-competent cells are selected against by this mechanism, while HGPRT-deficient cells survive.

Survival of such metabolically deficient cells is predicated on the existence of an alternate pathway by which guanine nucleotides are generated *de novo*. In fact, this is the prime pathway leading to purine nucleotides, while the HGPRT-mediated route is generally considered a salvage pathway of far less metabolic importance.

Once HGPRT-deficient mutants have been isolated by cloning in selection medium, revertants can be selected by taking advantage of another type of selection medium—in this case, medium containing thymidine (3 μg/ml), hypoxanthine (5 μg/ml), methotrexate (0.1 μg/ml), and glycine (7.5 μg/ml). We will henceforth and throughout this chapter refer to this medium as THMG selection medium. Selection against HGPRT-deficient mutants occurs in this medium because methotrexate is a powerful folic acid antagonist capable of preventing the *de novo* synthesis of purines and disrupting other single-carbon-transfer reactions that lead to the synthesis of thymidylate and certain amino acids. Provided the cell is HGPRT-deficient, the methotrexate-induced block proves lethal, and hence such cells are selected against. On the other hand, those cells with a wild-type allele at the HGPRT locus (the revertants) should survive THMG medium, since ribose phosphorylation of hypoxanthine and subsequent metabolism are sufficient to provide all the purine nucleotides required by the cells.

One other point, however, which needs mentioning is that cell survival also depends on the presence of an effective thymidine kinase. Without this enzyme, which phosphorylates thymidine, THMG medium would prove lethal, since the only other means of synthesizing thymidylate depends on the single-carbon-transfer reaction mentioned above, which is inhibited by methotrexate. In other words, and as becomes important to subsequent discussion, revertants of TK mutants as well as HGPRT revertants are positively selected for in THMG selection medium.

In initial attempts to isolate HGPRT mutants from L5178Y cells, we employed azg as a selective agent in the hope that all drug-resistant colonies would be amorphic or hypomorphic with respect to HGPRT activity. To our surprise, very high concentrations of azg were required to bring about appreciable killing of wild-type cells. As indicated in Table 1, concentrations of azg far in excess of those commonly employed in other cell systems (5–10 μg/ml) were needed. In fact, the upper concentrations employed approached within a factor of 2 the maximum solubility of the compound.

While it was possible to isolate a few mutant colonies from azg selection medium, a new problem was soon discovered. The proportion of cells killed by azg was found to be strongly dependent on the concentration

TABLE 1. Survival of HGPRT Wild-Type Cells in Azg-Containing Selection Medium

Azg (μg/ml)	Number of colonies/10–ml flask	Percent survival
0	69 ± 15	100
1.0	69 ± 9	100
3.2	64 ± 12	93
10.0	61 ± 7	88
32.0	8 ± 2	12
40.0	0	0

of cells plated. At high cell densities, a much greater proportion of cells survived than when lower cell concentrations were plated (Table 2). As illustrated by Table 2, the data on surviving cells as a function of plated cell concentrations are in accordance with a coincidence phenomenon. These observations have not been satisfactorily explained but could, of course, be predicated on the ability of the cells to detoxify the drug, i.e., metabolize azg in such fashion that further conversion to a purine nucleotide is either unlikely or impossible. This mechanism would also explain why such high concentrations of azg are required to induce death of wild-type cells.

Clearly, further efforts to exploit the use of azg as a selective agent seemed unjustified. It was therefore decided to test 6-mercaptopurine (6-MP), since this substance is also known to undergo ribose phosphorylation through the action of HGPRT enzyme. After several dose–response experiments and some consternation, a standard concentration of 50 μg/ml of 6-MP was deemed suitable. As indicated by the data presented in Table 3, the coincidence survival which troubled us when azg was employed was not a problem. We then proceeded to investigate the efficacy of 6-MP as a selective agent in a forward mutational assay system at the HGPRT locus.

Initially, it was necessary to know if the HGPRT mutants we had isolated and characterized (as regards the incorporation of ^{3}H-hypo-

TABLE 2. HGPRT⁻ Selection Medium (40 μg azg/ml): Coincidence Rescue

Number of HGPRT⁺ cells/10 ml	Number of "resistant" clones	Expected on basis of "coincidence"[a]	Apparent mutation frequency
$10^{4.0}$	9 ± 4	8	9×10^{-4}
$10^{4.5}$	69 ± 91	(69, assumed)	2×10^{-3}
$10^{5.0}$	708	625	7×10^{-3}

[a]Calculated from standard coincidence curves fitted to middle data point.

TABLE 3. Survival of HGPRT Wild-Type Cells in 6-MP-Containing Selection Medium

6–MP (μg/ml)	Number of cells plated	Colonies recovered
0	10^2	82 ± 5
50	10^4	0
50	10^5	0
50	10^6	0
50	10^7	0

xanthine) would be fully tolerant of 6-MP-containing selection medium. Satisfying ourselves that this was true, we had to determine whether the metabolic products of wild-type cells might not, upon lysis, contribute substrates which would prove lethal to the mutants. This problem was examined by making use of reconstruction experiments where HGPRT mutants were mixed with wild-type cells in known ratios prior to plating in 6-MP-containing selection medium. As the data of Table 4 indicate, 10^2 mutants per flask were readily recovered as colonies in the presence of 10^4 wild-type cells, which eventually lysed in 6-MP selective medium. However, when 3×10^4 wild-type cells were incorporated in the reconstruction with 10^2 mutants, mutant recovery, as evidenced by colonies, was depressed by a factor of 2. When 10^5 wild-type cells were employed against 10^2 mutants, no colonies were observed.

Presumably, the ability of wild-type cells to interfere with colony development among mutants is attributable to the cross-feeding of ribose-phosphorylated 6-MP from wild-type to mutant cells. Regardless of the precise reason, it is evident that use of inconveniently low cell densities is obligatory. For instance, to derive statistically valid information regarding mutant frequencies ranging in the order of 10^{-6}, approximately 300 Falcon flasks or 3 liters of cloning medium would be needed. For this reason, we have developed a method employing a different type of flask, into which 1 liter of cloning medium can be added. This approach, described elsewhere in greater detail (Appendix I, Section I), lends itself well to identification of mutant colonies under circumstances of low mutant frequencies.

Nevertheless, determining the spontaneous mutation rate at the HGPRT locus following a single generation in nonselective medium would be extremely difficult, requiring 30 or more large flasks. Based on mutant accumulation after many generations in nonselective medium, we have estimated the spontaneous mutation rate to be of the order of 10^{-7} mutation/locus/generation.

TABLE 4. HGPRT⁻ Selection Medium (50 μg 6-MP/ml): Reconstruction

Number of HGPRT⁺ cells/10 ml	Number of HGPRT⁻ cells/10 ml	Normalized recovery of HGPRT⁻ cells ($\pm 1\ \sigma$)
0	10^2	100 ± 9
$10^{4.0}$	0	0
$10^{4.0}$	10^2	91 ± 13
$10^{4.5}$	0	0
$10^{4.5}$	10^2	46 ± 11
$10^{5.0}$	0	0
$10^{5.0}$	10^2	0

Because of the necessity of working at low cell densities, it is clearly indicated that attempts be made to exploit other loci if the advantages of the L5178Y line are to be utilized. With this purpose in mind, work on the TK locus was initiated.

III. TK LOCUS: GENERAL PRINCIPLES

Thymidine monophosphate (TMP) occupies a unique position in DNA replication. Of the four principal deoxyribonucleoside monophosphates, it alone does not undergo significant conversion to other nucleotides. Probably as a consequence of this conservation, the TMP pool size remains fairly constant and small under normal growth conditions (Flamm, unpublished results) and serves to regulate DNA synthesis. These properties of small size and ability to arrest DNA synthesis (and hence cell growth) upon depletion make the TMP pool a highly sensitive target for selective agents in a mutational assay system. By focusing on these pool properties, one can kill a cell in two distinct fashions. First, because of the small pool size, one can kill by replacing most of the TMP with a lethal TMP analogue. An example of such an analogue is 5-bromodeoxyuridine (BUdR) monophosphate, which kills following incorporation into DNA. BUdR monophosphate results from the phosphorylation of BUdR by the "salvage" enzyme thymidine kinase (TK) (Fig. 1), which normally phosphorylates thymidine to TMP in most mammalian cells. TK-deficient mutant cells lack this enzyme activity and suffer no ill effects from BUdR treatment. BUdR, therefore, can select for TK-deficient cells and against TK-competent ones.

The second means by which one can lethally affect the TMP pool is by blocking TMP synthesis, thereby depleting the pool and arresting DNA synthesis and further cell growth. In TK-deficient cells, but not in TK-competent cells, there remains but a single pathway by which TMP is produced. In this pathway, the enzyme thymidylate synthetase (TS) methylates deoxyuridylic acid (dUMP) to TMP (Fig. 1). This TS methylation step requires continuous folate metabolism and hence can be blocked with the same THMG medium used for selection at the HGPRT locus. Whereas in TK-competent cells the TMP pool would be sustained by the phosphorylation of exogenous TdR by TK, in TK-deficient cells this is not possible, the TMP pool rapidly dwindles, DNA synthesis ceases, and the cells die. In this manner, THMG can select for TK-competent cells and against TK-deficient ones.

The effects of BUdR and THMG selection media on TK-competent ($TK^{+/+}$ or $TK^{+/-}$) and TK-deficient ($TK^{-/-}$) L5178Y cells are summarized in Table 5. (For reasons explained below, the TK locus is believed to be

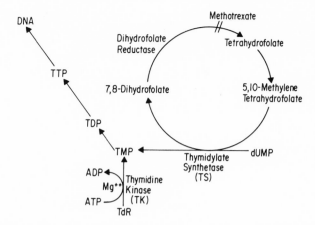

FIGURE 1. Biochemical pathways leading to the synthesis and utilization of thymidine monophosphate (TMP). TMP biosynthesis is effected both by the folate-dependent enzyme thymidylate synthetase (TS) and by thymidine kinase (TK). Using inhibitors of folate metabolism or lethal analogues of thymidine, the TS and TK pathways can be controlled to select for TK-competent or TK-deficient cells. For details, see text.

diploid. There are therefore two TK-competent genotypes—the $TK^{+/+}$ homozygote and the $TK^{+/-}$ heterozygote—and a single TK-deficient homozygote, $TK^{-/-}$.) The first two lines of Table 5 demonstrate the ability of $TK^{-/-}$ cells to clone with high efficiency in BUdR-supplemented cloning medium under conditions which select completely against $TK^{+/+}$ cells at 8×10^4 cells/ml plating concentration. When the two phenotypes are coplated at these same concentrations, there is only a slight, possibly insignificant depression of the $TK^{-/-}$ cloning efficiency resulting from the

TABLE 5. Reconstruction Experiments Using TK Homozygotes

Number of viable $TK^{+/+}$ cells per flask[a]	Number of viable $TK^{-/-}$ cells per flask[a]	Number of colonies in BUDR[b,d]	Number of colonies in THMG[c,d]
0	10^2	100 ± 9	—
8×10^5	0	0 ± 0	—
8×10^5	10^2	84 ± 9	—
10^2	0	—	100 ± 17
0	9×10^5	—	0 ± 1
10^2	9×10^5	—	104 ± 7

[a]10 ml of soft-agar cloning medium per flask.
[b]50 µg BUdR/ml soft-agar cloning medium.
[c]3.0 µg thymidine, 5.0 µg hypoxanthine, 0.1 µg methotrexate, and 7.5 µg glycine per milliliter of soft-agar cloning medium.
[d]Colony counts normalized relative to controls (100%); actual cloning efficiencies approximately 80% (mean ± 1σ).

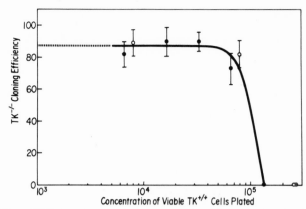

FIGURE 2.   Effect of $TK^{+/+}$ cell concentration on cloning efficiency of coplated $TK^{-/-}$ cells in BUdR selection medium (50 μg/ml). Ten $TK^{-/-}$ cells/ml were coplated with each of the indicated concentrations of $TK^{+/+}$ cells. The results are normalized relative to the cloning efficiency of $TK^{-/-}$ cells plated alone in nonselective cloning medium. Vertical bars represent standard deviations of replicate plates. Open and closed circles distinguish between two experiments performed several months apart.

metabolism and death of the $TK^{+/+}$ cells (line 3). The variation in $TK^{-/-}$ cloning efficiency with increasing concentrations of coplated $TK^{+/+}$ cells is explored further in Fig. 2. Only at cell concentrations above 8×10^4/ml is $TK^{-/-}$ mutant recovery affected.

The last three lines of Table 5 confirm in like manner the effectiveness of THMG as a selective agent for TK-competent cells in the presence of nearly 10^5/ml TK-deficient cells.

These two selection media provide the means for detecting both forward and reverse mutational events at the TK locus in L5178Y mouse lymphoma cells. We have subsequently shown that the mutation rate can be enhanced by known mutagens, that the genotypes and phenotypes are as ascribed—i.e., $TK^{+/+}$, $TK^{+/-}$, and $TK^{-/-}$—and that the system can be used for the detection of environmental mutagens.

IV.   TK LOCUS IN A MUTATIONAL ASSAY SYSTEM

The majority of L5178Y cells, when plated in THMG-supplemented cloning medium, develop as THMG-resistant clones. One of these clones, $TK^{+/+}-1$, was grown in suspension culture and served as the starting stock from which all subsequent mutant lines were ultimately derived. Table 6 confirms that this cell line is indeed TK competent.

The TK-deficient cell line, $TK^{-/-}-1.1$ (see Table 6 for TK activity), was grown from the single BUdR-resistant colony recovered after plating

TABLE 6.   Thymidine Kinase Activity of TK Mutants

Proposed genotype	TK activity[a]	Ratio[b]
$TK^{-/-}-1.1$	0.0	0.0
$TK^{+/-}-1.1.1$	25.6	0.93
$TK^{+/-}-1.1.3$	25.6	0.93
$TK^{+/-}-1.1.5$	27.5	1.00
$TK^{+/-}-1.1.6$	29.9	1.09
$TK^{+/-}-1.1.7$	28.5	1.04
$TK^{+/-}-1.1.9$	28.4	1.03
$TK^{+/-}-1.1.10$	29.2	1.06
$TK^{+/-}-1.1.12$	27.5	1.00
$TK^{+/-}-1.1.13$	26.5	0.96
$TK^{+/+}-1$	61.0	2.22

[a]Expressed as μμmoles of thymidine phosphorylated per 10^6 cells per minute of incubation.
[b]Based on setting the average of the $TK^{+/-}$ mutants equal to 1.00.

4.4×10^8 viable $TK^{+/+}$ cells, representing 2×10^{10} cells × generations (cells grown for approximately 50 generations in nonselective medium). The TK-competent and TK-deficient cell lines were used retrospectively in reconstruction experiments. With plating conditions identical to those used for the isolation of the $TK^{-/-}$ clone, a nearly quantitative recovery was observed for $TK^{-/-}$ cells (Table 5, Fig. 2). The observed forward mutation rate of 5×10^{-11} mutation/locus/generation appeared much too low to be attributed to a single genic event, so diploidy of the TK locus was invoked as a possible explanation. The assumption that the original THMG-resistant cell line and the BUdR-resistant cell line were homozygous (i.e., $TK^{+/+}$ and $TK^{-/-}$, respectively) led to the expectation that $TK^{+/-}$ heterozygotes would accumulate as spontaneous revertants in a $TK^{-/-}$ culture grown in nonselective medium ($TK^{-/-} \rightarrow TK^{+/-}$). The rate at which such spontaneous back-mutants accumulate should be greater than the figure of 5×10^{-11} mutation/locus/generation observed for the $TK^{+/+} \rightarrow TK^{-/-}$ event, since only one locus need revert to give rise to a TK-competent heterozygote. This was demonstrated by periodically determining the THMG-resistant mutant frequency of a $TK^{-/-}$ culture growing in nonselective medium for over 80 generations following removal from BUdR selection medium (Table 7). Note that while the mutant frequency increases with increasing time out of selection medium, the $TK^{-/-} \rightarrow TK^{+/-}$ mutation rate (frequency divided by number of generations) varies only randomly about a mean of 6×10^{-9} mutation/locus/generation.

A number of these presumed heterozygotes, as well as homozygotes, have been assayed for thymidine kinase activity. The results (Table 6)

TABLE 7. Spontaneous Back-Mutation Rate of $TK^{-/-}$ Cells in Nonselective Medium[a]

Number of generations in nonselective medium	Number of viable cells plated	Number of $TK^{+/-}$ mutants	Mutant frequency (mutations/locus)	Mutation rate (mutations/locus/generation)
6	0.75×10^8	3	4.0×10^{-8}	6.6×10^{-9}
34	4.80×10^8	88	18×10^{-8}	5.4×10^{-9}
46	0.80×10^8	15	19×10^{-8}	4.1×10^{-9}
48	0.65×10^8	30	46×10^{-8}	9.6×10^{-9}

[a]Total of 136 $TK^{+/-}$ mutants recovered. Mean ($\pm 1\sigma$) mutation rate equals $(6.4 \pm 2.4) \times 10^{-9}$ mutation/locus/generation.

are in excellent accord with the hypotheses that the TK locus is diploid (with no gene dose compensation mechanism) and that we have isolated the two homozygotes and a number of heterozygotes with our selection media.

Further support for denoting these cell lines as $TK^{+/+}$, $TK^{+/-}$, and $TK^{-/-}$ is provided by the spontaneous mutation rates of each of these genotypes to its opposite phenotype (i.e., THMG resistance $\longleftrightarrow$ BUdR resistance), as shown in Table 8. The determination of the $TK^{+/+} \rightarrow TK^{-/-}$ spontaneous mutation rate (approximately 5×10^{-11} mutation/locus/ generation) has already been described. Since it is based on the recovery of but a single mutant, the reliability of this figure is questionable; it probably represents a minimal figure. The $TK^{-/-} \rightarrow TK^{+/-}$ spontaneous reversion rate is based on the recovery of a sufficiently high number (well over 100) of mutants accumulated in nonselective medium over a large number of generations. We place high reliance on the figure of $(6.4 \pm 2.4) \times 10^{-9}$ mutation/locus/generation, assuming no selection for either pheno- type. Finally, the spontaneous $TK^{+/-} \rightarrow TK^{-/-}$ mutation rate has been determined by two independent means, namely, by the rate of mutant accumulation in a $TK^{+/-}$ culture over many generations and by the fluctuation test (Table 8). The two methods are in reasonable agreement, yielding an average spontaneous forward mutation rate $(TK^{+/-} \rightarrow TK^{-/-})$ of 1.2×10^{-7} mutation/locus/generation.

If we accept the above value for the $TK^{+/-} \rightarrow TK^{-/-}$ spontaneous mutation rate, we can calculate an expected $TK^{+/+} \rightarrow TK^{-/-}$ spontaneous rate. Assuming that two independent hits are involved $(TK^{+/+} \rightarrow TK^{+/-} \rightarrow TK^{-/-})$, spontaneous mutation rates of 2×10^{-7} and 1×10^{-7} mutation/ locus/generation, respectively, would be expected. (Since there are twice as many loci to be mutated to a defective state in the $TK^{+/+}$ homozygote as there are in the heterozygote, the expected rate for the first mutation should be double that observed for the heterozygote.) The overall $TK^{+/+}$

TABLE 8. Spontaneous Mutation Rates at the Thymidine Kinase (TK) Locus

Assumed mutational event	Phenotypic changes		Mutation rate (mutations/locus/generation)
$TK^{+/+} \rightarrow TK^{-/-}$	BUdR^s THMG^r	$\rightarrow$ BUdR^r THMG^s	Based on only 1 mutant (5×10^{-11})
$TK^{-/-} \rightarrow TK^{+/-}$	BUdR^r THMG^s	$\rightarrow$ BUdR^s THMG^r	$(6.4 \pm 2.4) \times 10^{-9}$
$TK^{+/-} \rightarrow TK^{-/-}$	BUdR^s THMG^r	$\rightarrow$ BUdR^r THMG^s	1.9×10^{-7} (rate of mutant accumulation in non- selective medium) 0.6×10^{-7} (fluctuation test) Mean $= 1.2 \times 10^{-7}$

$\rightarrow TK^{-/-}$ mutation rate should then be 2×10^{-14} mutation/locus/generation. This is clearly much lower than our observed rate based on a single $TK^{-/-}$ mutant (5×10^{-11} mutation/locus/generation, a factor of 2500 higher than expected), indicating considerable luck on our part in isolating the first all-important $TK^{-/-}$ mutant. Following our discussion of induced mutation rates, we will indicate how diploid loci in general might be treated to yield the corresponding homozygous recessive mutants in reasonable quantities for use in mutational assay systems.

In order to qualify as a mutational assay system, the $TK^{+/-} \rightarrow TK^{-/-}$ mutation rate must be subject to enhancement by a variety of mutagens. Since this mutational event is in the forward direction, one would not expect the mutagen specificity that one finds in a reversion system (Hartman *et al.*, 1972a; Capizzi *et al.*, 1973). Accordingly, we initially tested the system with two mutagens of widely divergent modes of action—X-irradiation and ethyl methanesulfonate (EMS)—and later with a suspected mutagen, hycanthone methanesulfonate (Hartman *et al.*, 1971b; Clive *et al.*, 1972a). The effects of such treatments are self-evident (Table 9). However, a word of explanation as regards the reporting of data is in order. By assuming that all induced genetic lesions were imparted within a single generation (11 hr for L5178Y cells *vs.* 2 hr or less of mutagen treatment), the observed mutant frequency per viable cell is numerically equal to an *induced mutation rate*. The fraction (induced mutation rate)/(spontaneous mutation rate), which we refer to as the induced rate factor (IRF), is a measure of the extent to which the "normal" or spontaneous rate of mutation has been enhanced by treatment with the agent in question. IRF values are tabulated for a few mutagen treatments in the last column of Table 9. Treatments with X-irradiation (2500-fold stimulation for 600 r), EMS (900-, 3000-, and 17,000-fold stimulations for 2 hr treatments at 10^{-3} M, $10^{-2.5}$ M, and 10^{-2} M, respectively), and hycanthone methanesulfonate [a linear response ranging from an 80-fold stimulation at 0.2×10^{-4} M (10 μg/ml) to a 450-fold stimulation at 1×10^{-4} M (50 μg/ml), all treatments being for 2 hr] consistently attest that the system responds to known and suspected mutagens in a dose-dependent manner.

One criticism of the L5178Y system might be that these cells are unusually mutable, possibly giving rise, for instance, to false positive results. Such an eventuality could have drastic economic effects on chemical and pharmaceutical industries if otherwise safe and efficacious substances were erroneously identified as mutagenic. Such complaints, however, have not been raised against the system employing the HGPRT locus in Chinese hamster cells. It might be informative at this point to examine data from the two systems before and after reduction to a common format. For example, a 2 hr exposure to 10^{-2} M EMS induced a 4.3-fold increase in *mutant frequency* at the HGPRT locus in Chinese

TABLE 9. Induced Forward Mutation Rates at the Thymidine Kinase (TK) Locus ($TK^{+/-} \rightarrow TK^{-/-}$)

Mutagenic agent	Dose	Relative survival (%)	Mutation rate (mutations/locus/generation)	IRF[a]
None (spontaneous)	—	100	1.2×10^{-7}	(0)
X-irradiation	600 r (136 r/min)	12	3×10^{-4}	2,500
EMS	$10^{-2.0}$ M, 2 hr	1.0	2×10^{-3}	17,000
EMS	$10^{-2.5}$ M, 2 hr	10	3×10^{-4}	3,000
EMS	$10^{-3.0}$ M, 2 hr	50	9×10^{-5}	900
Hycanthone	1×10^{-4} M, 2 hr	2	5.4×10^{-5}	450
Hycanthone	0.5×10^{-4} M, 2 hr	20	3.4×10^{-5}	280
Hycanthone	0.2×10^{-4} M, 2 hr	55	1.0×10^{-5}	80

[a] Induced rate factor equals $\dfrac{\text{Induced mutation rate}}{\text{Spontaneous mutation rate}}$.

Donald Clive, W. Gary Flamm, and James B. Patterson

TABLE 10. Relative Sensitivities of Chinese Hamster and L5178Y Mouse Lymphoma Cells to Mutagens

Mutagen	Cell type	Locus studied	Induced mutation rate[a]	Spontaneous mutation rate[a]	Induced rate factor	Reference
X-irradiation						
600 r	Chinese hamster	HGPRT[b]	$(1 \times 10^{-6}$ mutation/locus/r) 6×10^{-4}	1.5×10^{-8c}	40,000	Chu (1971b)
450 r	Chinese hamster	HGPRT[b]	$(1.8 \times 10^{-6}$ mutation/locus/r) 8×10^{-4}	$(1.5 \times 10^{-8})c$	50,000	Bridges and Huckle (1970)
600 r	Mouse (L5178Y)	TK	$(0.5 \times 10^{-6}$ mutation/locus/r) 3×10^{-4}	1.2×10^{-7}	2,500	Clive et al. (1972b)
EMS, 10^{-2} M, 2 hr	Chinese hamster	HGPRT[b]	32×10^{-5} ; 153×10^{-5}	1.5×10^{-8c}	20,000–100,000	Chu and Malling (1968a)
EMS, 10^{-2} M, 2 hr	Mouse (L5178Y)	TK	2×10^{-3}	1.2×10^{-7}	17,000	Clive et al. (1972b)

[a] Mutations/locus/generation.
[b] Mutation from 8-azaguanine sensitivity to 8-azaguanine resistance (30 µg/ml).
[c] From Chu et al. (1969).

hamster cells (Chu and Malling, 1968a); our system had the *mutation rate* stimulated 17,000-fold (Clive *et al.*, 1972b) over the spontaneous rate. Yet the induced *mutant frequencies*—the real effect—were remarkably close in the two systems, certainly within a factor of 10 at the outside. Table 10 compares our system with the Chinese hamster–HGPRT system as used in two laboratories (Chu and Malling, 1968a; Bridges and Huckle, 1970; Chu, 1971b). All of the data have been reduced to our manner of presentation, and induced rate factors have been calculated for each data set. It would appear from this table that the L5178Y mouse–TK system is the less sensitive of the two in response both to X-irradiation and to EMS. Even after assuming similar spontaneous mutation rates, the two systems do not differ significantly in their responses to similar mutagenic treatments.

It is most important—but outside the scope of this article—that such comparisons of spontaneous and induced mutation rates both *in vitro* and *in vivo* (e.g., Russell's, 1951, specific-locus system in mice) be extended. In this manner, one of the chasms separating *in vitro* mutagen assays and mutagenic effects in man (namely, the differences between the mutagen response of mammalian somatic cells *in vitro* and that of the whole mammal) can be bridged.

A final procedural point might be noted here. With a 2 hr exposure to 10^{-2} M EMS, a 17,000-fold stimulation of the spontaneous mutation rate is observed. A single such treatment of a $TK^{+/+}$ culture would yield a $TK^{+/-}$ mutant frequency of 4×10^{-3} (i.e., twice that frequency observed for $TK^{+/-} \rightarrow TK^{-/-}$ in Table 9). A second such treatment, acting on a $TK^{+/-}$ mutant frequency of 4×10^{-3} would yield a $TK^{-/-}$ mutant frequency of $(4 \times 10^{-3})(2 \times 10^{-3})$ or 8×10^{-6}, a frequency readily detectable by cloning in 50 µg 5-BUdR/ml. And, since the introduction of the second genetic lesion does not depend on prior expression of the first lesion, the same $TK^{-/-}$ mutant frequency should result if the two mutagen treatments are combined into a single treatment of 4 hr duration. This procedure could conceivably be extended to other autosomal loci once appropriate selective media have been developed.

V. CONCLUDING REMARKS

We have tried to indicate the utility of the thymidine kinase locus in mouse lymphoma cells and the advantages it holds over the more widely employed HGPRT locus. As indicated in this chapter, the amount of labor involved for the utilization of the TK locus is approximately one-tenth that for its X-linked counterpart. This advantage can be exploited to delineate slight increases in the mutation rate above the spontaneous level, providing not only the means of identifying weak mutagens but also the opportunity of characterizing possible threshold effects. A better understanding of the latter in mammalian somatic cells is of critical importance if we are to evaluate properly the existence of "no-effect" levels, should they in fact prove to be a reality.

As is true in many other scientific endeavors, the system described must be carefully monitored with all appropriate controls to assure that untoward events have not occurred. When the system works, it works well. But vigilance and a thorough understanding are obligatory to the attainment of meaningful results. The "*Mycoplasma* menace" is emphasized in Appendix II and constitutes one example of how results can go awry. By use of positive controls (where heterozygotes are mutagenized by a classical mutagen) and by including appropriate reconstruction experiments, one can be reasonably certain of the validity of the results obtained with test compounds.

Although the TK heterozygote in L5178Y cells has already proven to be of some utility, future developments are likely to greatly enhance its overall value. A continuing effort is planned to compare results obtained with different chemical mutagens to those derived from Russell's specific-locus test in whole mice. In such fashion, and with appropriate dosimetry, genetic events occurring in somatic cells can be correlated with those occurring in different germ-cell stages. Current efforts are designed to increase the versatility of the system to provide the means of delineating the molecular mechanisms by which mutations are induced at the TK locus. These will include the development of mutant stocks having well-defined genetic alterations. Attempts will also be made to incorporate the mutational assay system into the host-mediated assay concept. These efforts will involve the implantation of test cells in appropriate body cavities as well as the injection of such cells into general circulation.

Advances in mutation research, insofar as they concern problems to human health, have been greatly increased in the past few years, and it is evident that further progress is dependent on the continuing development of mutational assay systems having a high degree of relevance. It is hoped that this chapter will contribute to this necessary and urgent task.

VI. ACKNOWLEDGMENTS

We thank Miss Naomi Jean Bernheim and Mr. Michael R. Machesko for skilled technical assistance in the performance of the experiments reported here.

APPENDIX I.   CELL MATERIAL AND TECHNIQUES

A.   Cells

L5178Y mouse lymphoma cells (Fischer, 1958) are used throughout these studies.

B.   Media (Nonselective Suspension—Growth and Cloning)

L5178Y mouse lymphoma cells grow in suspension culture in Fischer's medium supplemented with penicillin (50 units/ml), streptomycin (50 µg/ml), and 10% horse serum (F_{10}), under a 5% CO_2-in-air atmosphere, with a generation time of 10.6–11.0 hr at 37°C. We find that the addition of sodium pyruvate (50 µg/ml) to F_{10} provides a medium (F_{10p}) which can support cell growth from a single-cell inoculum. From a one cell/ml inoculum, growth occurs with but a slightly increased generation time (12.0 hr), and this difference could be accounted for by statistical fluctuation in cell number when cultures are prepared at this low concentration. These media will be referred to henceforth as nonselective suspension media.

The cells clone with 80–100% efficiency relative to Coulter particle counts using a modification of the soft-agar cloning technique of Chu and Fischer (1968). Nonselective cloning medium (CM) consists of horse serum (20%), sodium pyruvate (50 µg/ml), and Noble agar (0.37%) in Fischer's medium containing the antibiotics. Since gelling of the CM is delayed for several hours at 37°C, there is ample time for taking aliquots of the medium, adding the appropriate concentration of cells, and mixing on a rotary shaker. Ten-milliliter fractions are pipetted into 30 ml plastic Falcon flasks in a sterile laminar flow hood, and the flasks are gassed with 5% CO_2-in-air, tightly capped, gelled in a –20°C freezer for 15 min, and incubated for 7–10 days at 37°C for colony growth.

C.   Mutant Stocks

The rationale for the selective media used in the isolation of the TK variants is discussed in Sections III and IV. Basically, the TK locus appears diploid, and the three possible genotypes—the $TK^{+/+}$ (TK-competent) and $TK^{-/-}$ (TK-deficient) homozygotes as well as a $TK^{+/-}$ (TK-competent) heterozygote—have been isolated and characterized.

D.   Selective Media (See Also Section III)

TK-competent cells ($TK^{+/+}$ and $TK^{+/-}$) grow in media supplemented with THMG (3.0 µg/ml thymidine, 5.0 µg/ml hypoxanthine, 0.1 µg/ml methotrexate, and 7.5 µg/ml glycine); these same media quantitatively kill TK-deficient cells.

TK-deficient cells ($TK^{-/-}$) grow in media supplemented with 50 µg/ml of 5-bromodeoxyuridine (BUdR); these media quantitatively kill TK-competent cells.

E. Thymidine Kinase Assay

TK activity is assayed by a modification of the method of Breslow and Goldsby (1969) and has been described in detail (Clive et al., 1972b).

F. Spontaneous Mutation Rates

Spontaneous mutation rates are determined by two independent means wherever possible: (1) the Luria and Delbrück (1943) fluctuation test using the P_0 method of Lea and Coulson (1949) and (2) the rate of mutant accumulation in nonselective medium over a large number of generations. Both methods have been discussed in detail elsewhere (Clive et al., 1972b). Table 8 lists our spontaneous mutation rates for both the $TK^{+/-} \rightarrow TK^{-/-}$ transition and the $TK^{-/-} \rightarrow TK^{+/-}$ reversion.

G. Maintaining Cultures with Low Mutant Background

Periodically, a small number of cells (100–3000, depending on the most recently observed mutant frequency) from a stock culture is added to 150 ml of F_{10p} and a culture is propagated from this inoculum. Assuming an unusually high mutant frequency of 10^{-5}, the use of even 3000 cells would allow but a 3% chance of transferring a mutant cell to the fresh medium. In practical terms, since mutant frequencies are generally less than 10^{-6}, the probability of such a transfer is only 0.3% or less. This mutant-free culture then acts as a single element in a fluctuation test and, in most instances, grows up to 5×10^5 cells/ml with a mutant frequency of about $1–2 \times 10^{-7}$ (see, e.g., Clive et al., 1972a, for typical background frequencies). From this point forward, the culture will accumulate mutants at the spontaneous rate based on statistical expectations.

The above technique works equally well for cultures of either phenotype. In addition to this dilution procedure, $TK^{-/-}$ cultures can be maintained free of revertants by culturing them in F_{10p} supplemented with BUdR (50 µg/ml). Two days prior to use, the cells are centrifuged and resuspended in fresh F_{10p} to rid the culture of residual BUdR.

THMG selection medium, although completely effective in cloning experiments, is not capable of suppressing the accumulation of mutants in $TK^{+/-}$ suspension cultures. This was first suspected when consistently high (approximately 10^{-5}) mutant frequencies were observed in several experiments shortly after removal of $TK^{+/-}$ cells from THMG. To confirm that spontaneously derived mutants were not being eliminated from a $TK^{+/-}$ culture in THMG-supplemented medium, the following experiment was performed. Using a 2 hr treatment with 3.2×10^{-3} M EMS, a high $TK^{-/-}$ mutant frequency was induced in each of two $TK^{+/-}$ cultures, one in F_{10p} supplemented with THMG and the other in F_{10p} alone. A third $TK^{+/-}$ culture in F_{10p} containing THMG was not mutagenized, but to it were added $TK^{-/-}$ cells to a final frequency of one $TK^{-/-}$ cell to every 10^4 $TK^{+/-}$ cells. At predetermined post-treatment times, a fraction of each of the cultures was removed from THMG-selective medium by centrifuging and resuspending the cells in F_{10p}. Two days later, these cells were plated in BUdR-supplemented cloning medium to assay for $TK^{-/-}$ mutant frequency. The results (Fig. 3) show that (1) $TK^{-/-}$ cells that had been added to the culture were rapidly eliminated in THMG-selection medium (over 100-fold decrease in $TK^{-/-}$ frequency within 24 hr); (2) the induced

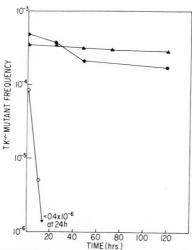

FIGURE 3. Ineffectiveness of THMG in selecting against newly arisen $TK^{-/-}$ mutants in a suspension culture of $TK^{+/-}$ cells. Fractions of a $TK^{+/-}$ culture were either (a) mutagenized with EMS to induce a high $TK^{-/-}$ mutant frequency, then maintained in F_{10p} (▲) or F_{10p} supplemented with THMG (●), or (b) supplemented with established stock $TK^{-/-}$ cells (final $TK^{-/-}$ frequency 10^{-4}) and maintained in THMG-supplemented F_{10p} (○). At the indicated times following treatment, aliquots were transferred to F_{10p} and, 2 days later, plated in BUdR-supplemented cloning medium for the determination of $TK^{-/-}$ mutant frequency.

$TK^{-/-}$ mutant frequency was not reduced but stayed constant at 3×10^{-4} for at least 120 hr in nonselective medium; and (3) apart from a two-fold decrease in the induced $TK^{-/-}$ mutant frequency during the initial 48 hr in THMG, no further significant effect of THMG on these induced mutants was observed during 120 hr of growth in the presence of THMG. Clearly, THMG does not effectively select against newly arising mutants in suspension culture.

Several of the 120 hr, THMG-resistant, induced-$TK^{-/-}$ mutant clones have been isolated and tested for methotrexate resistance. Roughly half of these mutants are resistant and half of them sensitive to 0.1 μg methotrexate per milliliter, suggesting that methotrexate resistance is partly but not entirely responsible for the persistence of mutants in THMG. For these reasons, we no longer rely on THMG-supplemented suspension medium for maintaining low mutant backgrounds in $TK^{+/-}$ cultures but utilize the previously described dilution method instead.

H. Mutagen Assay (Forward)

A typical mutagen assay experiment proceeds as follows:

1.   Approximately 1 week prior to the experiment, a $TK^{+/-}$ culture is started from a small (100–3000 cells) inoculum.

2.   One week later, the cells have grown to full culture size (300 ml at 5×10^5 cells/ml). Subcultures are prepared from it, one or more for each concentration of the compound to be tested, including untreated controls. All cultures are adjusted to a cell concentration of 3×10^5/ml.

3.   The compound under scrutiny is added from a freshly prepared stock solution to the required final concentration. Each culture is regassed and incubated with shaking for 2 hr, then centrifuged and washed twice in Fischer's medium lacking horse serum (same osmotic properties). The final washed cells are resuspended in F_{10p} for continued growth to permit expression of any genetic lesions that might have resulted from treatment.

4.   During the expression period, records pertaining to the dilution of cell cultures with fresh medium are maintained to permit accurate assessment of each culture's growth rate and hence an estimate of viability relative to a control culture.

5.   Optimal expression time is generally 48 hr (about four generations), but it may vary markedly from one mutagen to another. Following expression, 1.6×10^7 cells are centrifuged from each culture and resuspended in 200 ml CM (final concentration of 0.8×10^5 Coulter particle counts per milliliter CM). From this point onward, all cultures are handled identically so that no distinction will be made between control and treated.

6.   A typical mutagen assay experiment is comprised of four groups of plated Falcon flasks referred to below as series A through D.

a.   In order to assess cell viability, a 0.5 ml sample of the above cell suspension is diluted into 50 ml CM (approximately 10^3 cells/ml) and mixed thoroughly. A further 0.5:50 dilution is made, and 10-ml aliquots are pipetted into Falcon flasks (approximately 10^2 cells per Falcon flask) (series A.)

b.   For later use in reconstruction, $TK^{-/-}$ cells are pipetted into CM to a final concentration of 10^3 cells/ml. From this suspension, a $10^1/$ ml dilution is made in CM containing 50 μg BUdR/ml. Four 10-ml fractions of this dilution are plated in Falcon flasks in order to assess the cloning efficiency of $TK^{-/-}$ cells in selection medium (series B). (The 10^3/ml dilution is saved for use in series D below.)

c.   BUdR (final concentration 50 μg/ml) is added from a fresh stock solution to the remaining cells in approximately 200 ml CM. Fifteen 10-ml fractions are plated in Falcon flasks (0.8×10^5 cells/ml) for the determination of mutant frequency (series C). To the 50 ml of cell suspension which remains, 0.5 ml of 10^3 $TK^{-/-}$ cells/ml is added and four 10-ml fractions are plated in Falcon flasks for reconstruction (series D).

d.   All Falcon flasks are gassed, tightly capped, gelled for 15 min at $-20°C$, then incubated for 7–10 days at 37°C to permit colony growth. The flasks are then scored, and the number of colonies per Falcon flask is recorded.

7.   To calculate the induced mutation rate, let us assume the following colony counts:

Series A (treated):   40 clones/Falcon flask; hence the total number of viable plated cells in treated series C is 4×10^5/Falcon flask.

Series B: 90 clones/Falcon flask; hence 90 $TK^{-/-}$ cells are added to each Falcon flask in both treated and untreated series D.

Series C (treated):   60 clones/Falcon flask equals 60 mutants/4×10^5 cells equals 1.5×10^{-4} induced mutant frequency.

Series A (untreated control):   80 clones/Falcon flask; hence the total number of viable plated cells in untreated series C is 8×10^5/Falcon flask.

Series C (untreated control):   0.6 clone/Falcon flask equals 0.6 mutant/8×10^5 cells equals 8×10^{-7} mutant frequency in untreated control. Recovered induced mutant frequency equals 1.5×10^{-4} minus 8×10^{-7} equals 1.5×10^{-4}.

Series D (treated):   130 clones/Falcon flask. This should equal the sum of series B and C (i.e., 90 plus 60 or 150 clones) for 100% recovery of mutants; in this instance, it represents (130/150 × 100 or 87% recovery. The observed induced mutant frequency then is only 87% of the actual figure. The corrected induced mutant frequency becomes 1.7×10^{-4}. By assuming that these lesions are introduced within a single generation, we can claim an induced mutation *rate* of 1.7×10^{-4} mutation/1 locus/generation. Dividing this figure by the spontaneous rate of 1×10^{-7} mutation/locus/generation, we can assert that the given treatment has stimulated the spontaneous mutation rate 1700-fold, or that the induced rate factor is 1700 (see Section IV).

I. MUTAGEN ASSAY (REVERSE)

The only change in technique between the forward and reverse mutational assay systems stems from the 20-fold difference in the spontaneous rates of the $TK^{+/-} \rightarrow TK^{-/-}$ and the $TK^{-/-} \rightarrow TK^{+/-}$ mutations. Thus, instead of plating 150 ml of $TK^{+/-}$ cells in BUdR-supplemented CM (10 ml/Falcon flask), approximately 3000 ml of $TK^{-/-}$ cells are cloned in THMG-supplemented CM (10^5 cells/ml), 1000 ml per low-form (2800 ml) culture flask. Calculations are adjusted to allow for this difference in total plated cell number.

APPENDIX II.   THE *Mycoplasma* MENACE

Stanbridge (1971) has thoroughly reviewed the problems intrinsic to mycoplasmic contamination of cell cultures. Following such contamination we have found that our cultures acquire a number of unusual traits, most of which are predictable in accordance with Stanbridge (1971). Briefly stated, these are the following:

1.   Infected TK-competent cultures fail to grow in THMG-supplemented medium but continue to grow well indefinitely in unsupplemented medium.

2.   Infected TK-competent cells fail to remove ^{3}H-TdR at low concentrations from the growth medium (F_{10p}), whereas noninfected TK-competent cells rapidly incorporate ^{3}H-TdR. This inability to utilize exogenous thymidine serves to explain the failure of infected cultures to grow in THMG medium.

3.   Infected TK-competent cultures have normal TK activity in cell-free extracts, implying that their inability to take up low concentrations of thymidine from the medium results either from a transport problem or from an interaction between the PPLO and the added thymidine.

4.   Infected $TK^{+/-}$ cultures cannot be "mutagenized"; i.e., no BUdR-resistant colonies can be recovered from such cultures either

spontaneously or following mutagen treatment (for explanation, see point 8).

5. The bulk of the PPLO are probably in the supernate. The low-speed supernatant of infected TK-competent or TK-deficient cultures, when mixed with a low-speed pellet of noninfected TK-competent cells, completely inhibits ³H-TdR uptake from the culture medium by these cells. Such does not occur when a supernatant from a noninfected culture is mixed with noninfected TK-competent cells. Further, a contaminated supernatant can be freed of PPLO by a 30 min centrifugation at 50,000 × g so that it no longer interferes with ³H-TdR uptake. The pelleted material is morphologically identical with previously published electron microscopic studies on *Mycoplasma*.

6. In electron micrographs, PPLO appear both extracellularly and intracellularly.

7. The PPLO in our cultures can incorporate thymidine (hence BUdR) into their DNA. This was demonstrated by incubating $TK^{-/-}$ lymphoma cells with the supernatant from an infected culture in the presence of 20 µc/ml ³H-thymidine for 6 hr, pelleting the culture at 50,000 × g for 30 min, extracting the DNA, and banding it in a neutral cesium chloride gradient. The results are shown in Fig. 4. The tritium peaks to the light side of the mouse satellite DNA ($\rho = 1.691$ g cm⁻³) at an estimated density of 1.684 g cm⁻³, a figure in close agreement with published *Mycoplasma* DNA density of 1.686 g cm⁻³ (Sober, 1970).

8. $TK^{-/-}$ cultures can be maintained PPLO free by growth in 5-BUdR-supplemented medium. This follows from the fact that the PPLO possess TK activity and can lethally phosphorylate BUdR, while TK-

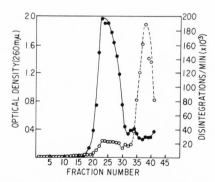

FIGURE 4. Density gradient analysis of DNA extracted from a PPLO-infected $TK^{-/-}$ culture grown in the presence of 20 µc/ml of ³H-thymidine (6.7 c/mmole). The gradient was centrifuged to equilibrium in a CsCl solution having an initial density of 1.705 g cm⁻³. Optical densities (●—●) were recorded in a Beckman DB spectrophotometer, and radioactivity (○—○) was assessed in a Packard Tri-Carb scintillation spectrometer.

deficient cells cannot. If high PPLO concentrations exist, cross-feeding and killing of the $TK^{-/-}$ cells are possible, especially in the event of intracellular PPLO. Evidence for this is shown in Fig. 4 by the radioactivity present in the mouse DNA. Such cross-feeding explains why $TK^{-/-}$ mutants are not recovered (in BUdR-supplemented cloning medium) from a mutagenized $TK^{+/-}$ culture (see point 4, above).

9. Contrariwise, infected $TK^{-/-}$ cultures cannot be induced to yield $TK^{+/-}$ revertants in THMG-supplemented cloning medium. This is a result of the inability of PPLO-infected TK-competent cells to take up thymidine from THMG-supplemented medium.

Attempts have been made to eliminate PPLO from cultures using a high level of kanamycin (400 µg/ml), which suppresses growth of PPLO and yields cell cultures which will take up thymidine from the medium. However, all such cultures develop the symptoms of PPLO infection soon after removal from kanamycin. To date, only high levels (50 µg/ml) of BUdR have routinely succeeded in maintaining $TK^{-/-}$ cultures free of PPLO. Aliquots of all $TK^{+/-}$ cultures are preserved in liquid nitrogen. Other aliquots are maintained in THMG-supplemented medium to serve as back-up cultures; such a culture is considered PPLO free if it continues growing in this medium.

VII. REFERENCES

Breslow, R. E., and Goldsby, R. A. (1969), Isolation and characterization of thymidine transport mutants of Chinese hamster cells, *Exptl. Cell Res. 55*, 339–346.

Bridges, B. A., and Huckle, J. (1970), Mutagenesis of cultured mammalian cells by X-radiation and ultraviolet light, *Mutation Res. 10*, 141–151.

Capizzi, R. L., Smith, W. J., Field, R., and Papirmeister, B. (1973), A host-mediated assay for chemical mutagens using the L5178Y/Asn⁻ murine leukemia, *Mutation Res. 21*, 6.

Chu, E. H. Y. (1971a), Induction and analysis of gene mutations in mammalian cells in culture, *in* "Chemical Mutagens: Principles and Methods for Their Detection" (A. Hollaender, ed.), Vol. 2, pp. 411–444, Plenum Press, New York, London.

Chu, E. H. Y. (1971b), Mammalian cell genetics III. Characterization of X-ray-induced forward mutations in Chinese hamster cell cultures, *Mutation Res. 11*, 23–34.

Chu, E. H. Y., Brimer, P., Jacobson, K. B., and Merriam, E. V. (1969), Mammalian cell genetics I. Selection and characterization of mutations auxotrophic for L-glutamine or resistant to 8-azaguanine in Chinese hamster cells *in vitro*, *Genetics 62*, 359–377.

Chu, E. H. Y., and Malling, H. V. (1968a), Mammalian cell genetics II. Chemical induction of specific locus mutations in Chinese hamster cells *in vitro*, *Proc. Natl. Acad. Sci. 61*, 1306–1312

Chu, E. H. Y., and Malling, H. V. (1968b), Chemical mutagenesis in Chinese hamster cells *in vitro*, *Proc. Twelfth Internat. Congr. Genet. 1*, 102.

Chu, M.-Y., and Fischer, G. A. (1968), The incorporation of ³H-cytosine arabinoside and its effect on murine leukemic cells (L5178Y), *Biochem. Pharmacol. 17*, 753–767.

Clive, D., Flamm, W. G., and Machesko, M. R. (1972a), Mutagenicity of hycanthone in mammalian cells, *Mutation Res. 14*, 262–264.

Clive, D., Flamm, W. G., Machesko, M. R., and Bernheim, N. J. (1972b), A mutational assay system using the thymidine kinase locus in mouse lymphoma cells, *Mutation Res. 16*, 77–87.

Davidson, J. D., Bradley, T. R., Roosa, R. A., and Law, L. W. (1962), Purine nucleotide pyrophosphorylases in 8-azaguanine-sensitive and -resistant P388 leukemias, *J. Natl. Cancer Inst. 29*, 789–803.

Fischer, G. A. (1958), Studies of the culture of leukemic cells *in vitro, Ann. N.Y. Acad. Sci. 76*, 673–680.

Hartman, P. E., Hartman, Z., Stahl, R. C., and Ames, B. N. (1971a), Classification and mapping of spontaneous and induced mutations in the histidine operon of *Salmonella, Advan. Genet. 16*, 1–34.

Hartman, P. E., Levine, K., Hartman, Z., and Berger, H. (1971b), Hycanthone: A frameshift mutagen, *Science 172*, 1058–1060.

Kao, F. T., and Puck, T. T. (1968), Genetics of somatic mammalian cells, VII. Induction and isolation of nutritional mutants in Chinese hamster cells, *Proc. Natl. Acad. Sci. 60*, 1275–1281.

Lea, D. E., and Coulson, C. A. (1949), The distribution of the numbers of mutants in bacterial populations, *J. Genet. 49*, 264–285.

Lesch, M., and Nyhan, W. L. (1964), A familial disorder of uric acid metabolism and central nervous system function, *Am. J. Med. 36*, 561–570.

Lieberman, I., and Ove, P. (1960), Enzyme studies with mutant mammalian cells, *J. Biol. Chem. 235*, 1765–1768.

Littlefield, J. W. (1963), The inosinic acid pyrophosphorylase activity of mouse fibroblasts partially resistant to 8-azaguanine, *Proc. Natl. Acad. Sci. 50*, 568–576.

Luria, S. E., and Delbrück, M. (1943), Mutations of bacteria from virus sensitivity to virus resistance, *Genetics 28*, 491–511.

Morrow, K. J., Jr. (1964), An investigation into the basis of variation of purine analogue resistance in an established murine cell line, Ph. D. thesis, University of Washington.

Morrow, K. J., Jr. (1970), Genetic analysis of azaguanine resistance in an established mouse cell line, *Genetics 65*, 279–287.

Russell, W. L. (1951), X-ray-induced mutations in mice, *Cold Spring Harbor Symp. Quant. Biol. 16*, 327–336.

Seegmiller, J. E., Rosenbloom, F. M., and Kelley, W. N. (1967), Enzyme defect associated with a sex-linked human neurological disorder and excessive purine synthesis, *Science 155*, 1682–1684.

Sober, H. A. (ed.) (1970), "Handbook of Biochemistry: Selected Data for Molecular Biology," 2nd ed., Chemical Rubber Co., Cleveland.

Stanbridge, E. (1971), Mycoplasmas and cell cultures, *Bacteriol. Rev. 35*, 206–227.

Szybalski, W., and Szybalska, E. H. (1962), Drug sensitivity as a genetic marker for human cell lines, *Univ. Mich. Med. Bull. 28*, 277–293.

Szybalski, W., Ragni, G., and Cohn, N. K. (1964), Mutagenic response of human somatic cell lines, *Symp. Internat. Soc. Cell Biol. 3*, 209–221.

CHAPTER 28

Approaches to Monitoring Human Populations for Mutation Rates and Genetic Disease

J. V. Neel

The University of Michigan Medical School
Ann Arbor, Michigan

and

T. O. Tiffany and N. G. Anderson

Molecular Anatomy Program*
Oak Ridge National Laboratory†
Oak Ridge, Tennessee

I. INTRODUCTION

Anxiety concerning the genetic effects of chemical mutagens in the environment and in food, of radiation, and of infectious agents is widespread. It is the responsibility of human geneticists and those with allied interests to obtain hard data on these effects so that wise decisions can be made and sensible actions taken. It is not sufficient to limit studies to experimental animals and then extrapolate the results to man. Species may differ in their susceptibilities to mutagenic agents, rates and mechanisms

*The Molecular Anatomy Program is sponsored by the National Institute of General Medical Sciences, the National Cancer Institute, the National Institute of Allergy and Infectious Diseases, and the U.S. Atomic Energy Commission.
†Operated for the U.S. Atomic Energy Commission by the Nuclear Division of Union Carbide Corporation.

of repair may differ among species, and the data on experimental organisms may be obtained only years after human exposures have been initiated. While model studies must be expanded and extended, the overriding concern is with man himself, and with the effect of current insults. Hence every effort must be expended to obtain reliable human data. No experimental animal population can be a satisfactory indicator in real time of the sum total of the panoply of potential mutagens to which man is subject.

The study of both spontaneous and induced mutation in man is in a generally unsatisfactory state. In this chapter, we adopt the view that ultimately the detection of human mutations by biochemical methods will be done on a large scale, that it will be integrated into our evolving health care system, and that many of the same measurements will provide data important to the geneticist and of value to the physician and the medical statistician. While emphasis here is on genetic screening, we cannot overlook the possible or probable additional medical implications of many of the findings. We stress the initial use of one approach, based on electrophoretic variants. An ongoing program must gradually adopt a variety of techniques, however, as these are adapted to the screening laboratory. For more general reviews of most of the possible approaches to monitoring for an increase in human mutation rates, the reader is referred to Neel and Bloom (1969), Vogel (1970), Crow (1971), and Neel (1970, 1971). We will not be concerned here with the subject of mutations involving whole chromosomes or parts thereof (see Cohen and Hirschhorn, 1971; Cohen and Bloom, 1971; Shaw, 1972), except to note that monitoring for chromosomal mutations can be readily combined with the systems described here when the cytological techniques and computerized pattern recognition systems have been developed to the point where they yield highly reproducible results under screening conditions.

II. THEORETICAL IMPORTANCE OF MEASUREMENTS IN MAN

The approach to be advocated for initial use requires a major effort, at a considerable expense. It may be justified on both theoretical and practical grounds. Most of the contributions to these volumes have assumed that for the test organism under consideration there are satisfactory methods for evaluating spontaneous rates of mutation; these have concentrated on how to detect significant increases above the spontaneous, "normal" rate. For man, as is well known, our most valid estimates of mutation rates are based on the frequency of occurrence of isolated cases of certain phenotypes which, once they appear in a kindred, are transmitted as dominant traits. Since one elects to perform a mutation rate study only if the phenotype in question occurs with a "noteworthy" frequency, there is obvious selection involved in the traits studied. Furthermore, in man it has never been possible to conduct the kind of genetic analysis which enables one to conclude whether the phenotype in question may result from mutation at two or more different genetic loci. The average of the apparently satisfactory mutation rate estimates in man is approximately 2×10^{-5}/locus/generation (cf. Vogel, 1970; Neel, 1971). Because of the possible bias introduced by a few relatively high estimates (multiple neurofibromatosis, polycystic disease of the kidney), a figure of 1×10^{-5} is often used in calculations. Stevenson and Kerr (1967), considering sex-linked traits, feel that a better overall estimate is 1×10^{-6}, a viewpoint endorsed by Cavalli-Sforza and Bodmer (1971). For all these estimates, the proportion of the mutational spectrum of that locus being detected is unknown. Neel (1957, et seq.) has emphasized the possibility that for some loci a much higher proportion of mutations may have gross phenotypic effects than for others, in which case the higher rates would not be so biased as appears at first thought.

Despite these obvious sources of bias, the average of the estimates available for man does not differ greatly from the presumably less biased estimates for the mouse (Russell, 1965; Schlager and Dickie, 1966; Lyon and Morris, 1966; Schlager, 1972; see also Lüning and Searle, 1971) or, for that matter, those for *Drosophila*, the average being about 0.8×10^{-5} for both organisms (but note the absence thus far of mutations in the control series of Lyon and Morris, 1966). However, whereas for man the estimates are based on autosomal dominant and sex-linked and recessive mutations compatible with survival, for the latter two species the estimates derive from the use of special stocks in which for the most part "recessive" autosomal mutations are being detected; for the mouse, the estimate for dominants alone is 0.4×10^{-6}. Now we must ask whether the mice were scrutinized with the same attention that people generally receive.

Numerous investigators have commented from time to time on the apparent general similarity of these rates, despite the many differences in the three species involved, and have speculated as to what extent the similarities are inherent in certain aspects of the detection process. This problem of apparent similarities in mutation rates among very different organisms persists even if we accept the thesis that existing estimates are highly biased and that a more reasonable figure is 1×10^{-6}/gene/generation. A strong challenge to the validity of these apparent similarities has now arisen from recent theoretical developments. Kimura and colleagues (1968a, b, 1969, et seq.; see also King and Jukes, 1969), on the basis of a consideration of the amino acid composition of some seven different proteins of widespread occurrence in both the plant and animal kingdoms, have concluded that the rate of substitution of one amino acid for another in the polypeptide chain has been approximately (and impressively) constant with time in all lines of descent. It is argued that this observation is inconsistent with what would be expected if the substitutions were due to natural selection, but it poses no problem if the substituted amino acid is functionally equivalent to its predecessor, the substitution then being due to the chance fixation of a mutation whose corresponding phenotype is neutral. While there is no doubt that some mutations result in phenotypes subject to positive or negative selection, Kimura and Ohta (1971, p. 28) "believe the conclusion is inevitable that in the human genome, nucleotide substitution has an appreciable effect on fitness in only a small fraction of DNA sites, possibly less than one percent of the total."

Not unexpectedly, this bold speculation has aroused intense controversy. Many of the known (or probable) amino acid substitutions in enzymatic proteins are accompanied by quantitative alterations in the activity of the enzyme (review in Harris, 1971). While the alternative enzyme states thus far studied for the most part represent genetic polymorphisms, and so could be a selected group, if Kimura is correct and the present data are representative, this suggests a "looseness" in the organization of biological systems, an amount of "noise" in nature, which is quite inconsistent with the predominantly deterministic formulation of this century. Its implications for our view of biological man are far-reaching.

Evidence accumulated over the past several years, since Kimura enunciated this hypothesis, now makes it apparent that there have been real differences in the rate of amino acid substitutions in proteins (reviewed in Smith, 1970; Dickerson, 1971). But it could still be argued that there is "relative" constancy—whatever that means. In all modern genetics, there is probably no hypothesis more important to test than the nonadaptiveness of most nucleotide substitutions in evolution. And there is a consequence (or prediction) of this hypothesis which should be eminently testable. If mutation is constant with time, then on a per-generation basis there must be wide differences among organisms. During the past 4–5 million years, average generation time for the *Drosophila* line of descent has probably averaged 10–20 days, depending on temperature; for the mouse line of descent, 5–7 months; and for the human line, at least 10–15 years. This suggests mutation rates per generation in the ratio of 1:12:300, i.e., 2 orders of magnitude difference between *Drosophila* and man! This prediction cannot be adequately tested with the approaches presently available. How do we relate scarlet eye in *Drosophila* to short ear in the mouse to retinoblastoma in man? But if we compare the frequency of mutations involving a homologous protein in these three species, there is a high probability that we are considering the results of mutation at homologous genetic loci.

Very preliminary data from *Drosophila melanogaster* suggest that the average *detectable* mutation rate for some 13 allozyme loci is 4–5 x 10^{-6}/ gene/generation (Mukai, 1970; Tobari and Kojima, 1972). Several of the enzymes under study in *Drosophila* are similar in properties to those already investigated in studies on man (e.g., malate dehydrogenase, isocitric dehydrogenase), and the possibility of developing a roughly comparable battery for two species as unlike as *Drosophila* and man clearly exists. It should not be difficult to exclude for a man a rate 100 times the *Drosophila* rate, as predicted from the Kimura hypothesis. For instance, let us assume that the *detectable* mutation rate in man is 10^{-4}/gene/generation. If in man one conducted 40,000 locus tests, and found no mutations, the probability of this occurrence at that mutation rate is 0.018; i.e., one would already have data pertinent to this issue. Thus to the many long-standing "philosophical" reasons for desiring better knowledge of mutation rates in higher organisms, there has now been added a compelling new one.

III. PRACTICAL IMPORTANCE OF HUMAN GENETIC MONITORING

The practical need for the assessment of induced mutations in man continues to grow. In part, the problem is created by the increasing intrusion of ionizing radiation into our lives, as reviewed in the reports of several national and international committees (National Academy of Sciences, 1956; Medical Research Council, 1956, 1960; United Nations, 1962, 1969). These reports present in detail the amount and sources of radiation reaching the public and the nature of the deleterious effects to be expected. It is sufficient to note here that the matter of possible genetic risks continues to be the central issue in discussions of permissible exposure to radiation, and new debates erupt almost yearly.

In addition to concern over the genetic effects of ionizing radiation, the possibility of chemical mutagenesis—the subject of these volumes— is receiving increasing attention. There is little doubt that the various screening methods described in this series and elsewhere will detect the most potent mutagens, against which human populations should be carefully shielded. Generally overlooked is the possibility that "permissible" exposures to some 50 or 60 mild mutagens might cumulate, by either additive or synergistic effects, to produce a significant rise in mutation rates. This mixture of exposures could be approximated in the mouse, but the differences in metabolic pathways between primates and other animals and the cumulative effects of long primate life cycles must be considered. Quite obviously, data on man himself are not only highly desirable but may in fact ultimately be demanded by the public.

Thus there exist a multiplicity of valid reasons for desiring valid data on mutation frequency in man on a continuing basis. The question then becomes one of how these measurements should or could be made.

IV. STUDY OF MUTATION AT THE MOLECULAR LEVEL

In theory, it is possible to screen for mutation at six levels:

1. Base changes in DNA.
2. Base changes in RNA.
3. Changes in amino acid sequence of proteins (primary structure).
4. Changes in protein function, measured directly by determining enzyme activity or antigenicity (secondary and tertiary structure).
5. Alterations in the level of enzymes or enzyme products, i.e., metabolites.
6. Morphological or behavioral changes (gross phenotypic changes).

In the past, the sixth approach has been the one pursued for higher organisms. The methodology and findings have been frequently reviewed and need not detain us here. The other approaches are relatively new. The first two are not technically feasible at present, but the third through fifth are. Judicious use of the third permits rather strong inference concerning events at the level envisaged by approaches 1 and 2.

At present, the approach of preference appears to be the search for qualitative differences, such as the appearance of an enzyme band in an electropherogram where none was present before, rather than a quantitative difference in a substance which is always present. For normal blood or urine constituents, normal values vary over a considerable range, and the values seen in a subpopulation known to have a given genetic disease may overlap it. This is especially true for metabolites, where alternate biochemical pathways and compensatory changes due to either homeostatic mechanisms or therapy may exist. Hence our initial emphasis is on the detection of qualitative changes in substances which are direct gene products.

The approach advocated in this presentation for initial use is conceptually and technically simple. Electrophoresis has been shown to be a powerful tool in detecting those amino acid substitutions in proteins which are accompanied by a change in molecular charge under specified conditions. As of this writing, the technique has led to the recognition of qualitative variants with respect to some 40 different protein components of human blood serum or erythrocytes. The number is constantly expanding. It is suggested that a suitable subset of these could be utilized in a study of spontaneous mutation and a search for evidences of an increased mutation rate in defined exposed populations. Conceptually, study of spontaneous rates and study of induced rates go hand in hand. Blood specimens are drawn and examined for the occurrence of variant proteins. In the present state of knowledge, it appears preferable to concentrate on proteins with respect to which variants are not known to occur in polymorphic proportions, since otherwise many rare variants with the same electrophoretic mobility as a well-known variant will be passed over. When a variant is found, further specimens are obtained from the father and mother of the variant individual. If the abnormality is present in one or the other of these persons, then the variant is known to be inherited. But if neither parent is affected, the variant *may* be due to mutation. The most likely alternative explanation is a discrepancy between legal and biological parentage, which can with present procedures be detected in most instances (see below).

Not all of the protein constituents of the blood serum and erythrocytes of man will be equally suitable for studies of mutation rates. The proteins (including enzymes) employed for this purpose should be relatively stable, and variants should be readily identifiable. Table 1 presents a list of presently available proteins, classified according to present knowledge into four degrees of suitability. This list, a modification of an earlier list by Neel (1971), owes much to consultations with Drs. Harry Harris, Richard Tashian, and George Brewer. No effort will be made in this paper to discuss the current techniques of identifying variants in these systems; references are to be found in Giblett (1969), Brewer (1970), and Harris (1970). It is doubtful whether any two laboratories would produce quite the same list. However, it is clear that there are at this writing some 20 different proteins which seem adequate for this purpose. It is recognized that if the monitoring is done at birth, then some of the proteins on that list, such as carbonic anhydrase, would be inappropriate.

It would appear especially convenient in a monitoring program to work with cord blood from newborn infants. One establishes the desired sample size and the time interval over which the sample is to be accumulated. Presumably any monitoring program, once instituted, continues either indefinitely or until shown to be unnecessary. It would be well to decide in advance on the cycle of analysis—say every 5 years; this avoids the troublesome problems which arise when *a posteriori* decisions are made concerning the time intervals into which to break a continuous series.

TABLE 1. A Listing of Serum and Erythrocyte Proteins and Enzymes of Possible Usefulness in the Study of Mutation by Electrophoretic Techniques[a]

Class 1 (best)	Class 2	Class 3	Class 4 (worst)
		Serum	
1. Transferrin	1. α_1-Antitrypsin	1. α_2-Macroglobulin	1. Prealbumin
2. Albumin	2. Ceruloplasmin	2. Alkaline phosphatase (forward-migrating mutants only)	2. Pseudocholinesterase
		3. Haemopexin	
		Red cell	
1. Hemoglobin A ($\alpha\beta$)	1. Nonspecific "esterase"	1. Peptidase C	1. Hexokinase
2. Hemoglobin A_2 ($\alpha\delta$)	2. Isocitric dehydrogenase	2. Peptidase D	2. Serum amylase
3. Carbonic anhydrase I	3. Indophenol oxidase	3. Adenine phosphoribosyl transferase	3. Catalase
4. Carbonic anhydrase II	4. Glutamate oxalacetate transaminase	4. Nucleoside phosphorylase	4. Galactose-1-phosphate uridyl transferase
5. Adenosine deaminase	5. Phosphoglycerate kinase	5. Pyrophosphatase	5. Glyceraldehyde-3-phosphate dehydrogenase
6. Adenylate kinase			6. Leucine aminopeptidase
7. NADH-diaphorase			7. Aldolase
8. G-6-PD (X-linked)			8. Glutamate dehydrogenase
9. LDH ($\alpha\beta$)			9. Glutathione reductase
10. Malic dehydrogenase			10. Pyruvic kinase
11. Peptidase A			
12. Peptidase B			
13. 6-PGD			
14. PGM-1			
15. PGM-2			
16. Phosphohexose isomerase (RBC *or* serum)			
17. Glutamate pyruvic transaminase			

[a]The list is offered more as a point of departure for discussion than as a definitive compilation; it is certain to undergo rapid modification in the immediate future.

The question might be raised as to whether a variety of trace chemicals (plus radiation) have already effected a significant increase in mutation rates, i.e., whether current observations would be inadequate as a base line. Now, many of the exposures of concern are developments of the past generation. As an alternative to simply mounting an ongoing system of monitoring, one might examine systematically blood specimens from newborn infants and their mothers. If a variant is found in the mother, then her parents should be examined. With the average age at maternity between 25 and 30, one could examine the question of an increase over a 25- to 30-year period. This research design is somewhat more efficient than the preceding, since the two-generation data would automatically yield important information on the inheritance of any variants, but this gain would probably be offset by the greater difficulties in locating the parents of the mothers, when this became necessary. There is inherent in this approach the assumption that these mutations are selectively neutral in the heterozygote, i.e., that there is no relationship between presence of a mutation and likelihood of survival and reproduction.

V. CRITIQUE OF GENETIC MONITORING BASED ON ELECTROPHORESIS

There are a number of considerations which make it clear that the initial system proposed is less than perfect. These have recently been dealt with at some length (Neel, 1971; Weitkamp, 1971), and only a brief recapitulation is included here.

First, there is the problem of technical error, resulting in both "false positives" and "false negatives." Since one will never accept a variant as such without confirmation, "false positives" should be rare. "False negatives"—in this case, missed electrophoretic variants—are possible in any system. This would lead to an underestimation in mutation rates, but since in a monitoring system one is seeking for differences in rates of mutations in two populations, the essentials of successful monitoring will not be challenged as long as the proportion missed is the same in both populations (see below).

Second, discrepancies between legal and biological parentage introduce unavoidable "noise" into the system. With present techniques [including testing for the histocompatibility (HLA) antigens], in excess of 90% of such occurrences can be detected and, further, an estimate can be developed of the fraction of apparent mutations which must actually be attributed to undetected discrepancies.

Third, the occurrence of genetic polymorphisms with respect to some of the proteins which otherwise meet all of the criteria will detract from the usefulness of these particular indicators. In effect, this means that certain classes of forward or backward mutations cannot be detected. It would appear that suitable mathematical allowances can be made for this fact, but obviously for monitoring purposes it would be wise to avoid the polymorphic proteins insofar as possible. Whether this decision would introduce an element of bias with respect to the mutability of the proteins concerned is a moot question.

Perhaps the largest issue is the fourth one: What proportion of the mutational spectrum will be detected? Electrophoresis is the method *par excellence* for detecting the results of certain base-pair substitutions. However, in the case of the hemoglobin molecule, and presumably other molecules of similar complexity, roughly a third of the possible amino acid substitutions will result in those changes in molecular charge which result in clear electrophoretic differences. Again, to a large extent, allowance can be made for this in any calculation of mutation rates. Can mutations characterized by small, intracistronic duplications or deletions be detected? The experience with hemoglobin's $\beta^{Frieburg}$ and $\beta^{Gun\ Hill}$ reveals that deletions of one and five amino acids, respectively, can be detected (Jones *et al.*, 1966; Bradley *et al.*, 1967); the experience with hemoglobin β^{Tak} and with hemoglobin $\alpha^{Constant\ Spring}$ shows that the addition of ten and 33 residues, respectively, to the C-terminal end of the chain (with a number of possible explanations) can be demonstrated (Clegg *et al.*, 1971; Kinderlerer *et al.*, 1971). Possibly in a somewhat different category (should it be called a mutation?) are the eight different hemoglobin "Lepores," with the *N*-terminal portion characteristic of the α-polypeptide and the carboxyl terminal of the β-polypeptide, the interchange presumably due to asymmetrical crossing-over in the tandem duplication which codes for the α- and β-polypeptides of hemoglobins A_2 and A, respectively.

It seems apparent that some additions to, and subtractions from, polypeptide chains can be detected electrophoretically. Whether deletions or duplications involving an entire cistron can be detected is less clear. These would, of course, not be expected to produce the type of qualitative change detected by electrophoresis but should produce quantitative changes in enzyme activity. Controller-gene type mutations should also result in quantitative variations. It is doubtful that changes in the intensity of the staining reaction with most dye-reduction techniques would be a sufficiently reliable indicator of the occurrence of such mutations, but for certain selected enzymes techniques for quantitating enzyme levels might be applied to the detection of variants (see below).

A fifth weakness of the approach is the assumption that a child with a new amino acid substitution in a protein, whose other genetic characteristics are consistent with its parents, is in fact the result of a mutational event. This assumption is a standard problem in all studies of mutation rate and must be met in the usual way—by the demonstration that when such presumptively mutant individuals reproduce, they transmit the defect to half their children—even though it is difficult to conceive of another reason for an amino acid substitution in approximately half of the molecules of a given protein of an individual.

Finally, the absolute necessity of standardized techniques for a given time period must be mentioned. Once a monitoring program has been initiated, one must continue to utilize the same proteins (because mutation rates may vary from structural gene to structural gene) and to examine the protein the same way (because technical improvements could bring to light previously hidden variants) for the planned screening interval. Herein lies perhaps the greatest barrier to initiating a monitoring program immediately: the art of electrophoresis continues to evolve rapidly. The answer to this last problem is to add new procedures as they become available but retain a central core of approaches which will only be replaced after an overlapping trial with any battery of revised procedures.

The chief strength of an electrophoretic approach is the precision it brings to the study of one type of spontaneous and induced mutation in man. Granted that the entire mutational spectrum may not be adequately covered, for at least one portion coverage should be very good. No system of mutation detection in the past has begun to detect all the mutations at a particular locus for any organism, and surely data which can be equated to changes in the genetic code have much to offer over data on "dominant visibles," "recessive visibles," or "lethals."

A second equally important strength is the opportunity this approach provides to build bridges to experimental genetics. The difficulties which have arisen in attempting to extrapolate from the extensive investigations of murine radiation genetics to man require no elaboration. Given, however, one secure bridge, such as could be supplied by this approach, the problems of extrapolation should be substantially eased. If, for instance, despite the predictions of the Kimura formulation, the spontaneous rate of mutation to genes resulting in biochemical variants is the same in man and mouse, then one has more confidence in applying estimates of the "doubling dose" of a particular mutagen derived from mouse work to the human situation. For another example, if one is interested in estimating precisely the impact of a given mutagen (rather than in the "doubling dose" approach), one can establish ratios between mutations resulting in biochemical variants and gross phenotypic variants in the mouse; these ratios, once the rate for mutations resulting in biochemical variants is known for man, can probably be utilized to predict other types of genetic damage to our species.

Finally, possibilities for the study of somatic-cell mutation rates in man are emerging (*cf.* Sutton, 1971, 1972). If these methods could be substituted for studies of germinal-cell mutation rates, this would be highly desirable, since they will involve only a fraction of the effort and expense. An absolute *sine qua non* of any use of somatic-cell mutation rates to monitor human populations is that there has been at least one extensive study in which both somatic-cell and germinal-cell mutation rates have been studied simultaneously in the same population.

VI. DOES THE INITIAL APPROACH MEET THE CRITERIA FOR A SATISFACTORY MONITORING SYSTEM?

Crow (1971) has presented six criteria for a mutation-monitoring system, as follows:

1. Is the system relevant (i.e., does it tell us about the kind of genetic damage that is pertinent to human problems)?
2. How quickly will a mutation increase be detected?
3. Can the system detect a small increase in the mutation rate?
4. Can many kinds of mutational events be detected?
5. Does the system offer a high probability of identifying the cause of the mutation increase?
6. Is the system available now?

Criterion 5 strikes us as somewhat unrealistic in that, although epidemiological clues might arise were there, for example, a highly localized increase in mutation rates, from what is now known of the relative nonspecificity of mutagen effects it is unlikely that any monitoring system can be too helpful in implicating specific mutagens (although a biochemical system might be the best). This is scarcely a reason for not pressing the issue of an increase in mutation rates. The other five criteria are excellent. To these, however, we would add one more: Does the system tie in with experimental approaches, so as to strengthen the basis for extrapolation from laboratory animals to man? Finally, although it can scarcely be used as a criterion, we suggest that the question of whether the system advances our understanding of basic human genetics, so that possibly better systems can be derived, is an important one.

Judged by these criteria, mutation monitoring by biochemical techniques is surely the best approach available today. To us, the most important reservation as of now is whether it would detect the "important" mutations. It is certainly not detecting mutations with gross phenotypic effects. But if the qualitative enzymatic changes with quantitative effects are actually the "slightly deleterious" mutations which the *Drosophila* work suggests are so much more common than the clear-cut "visibles," then these may in fact be the more important class of mutations. Only an actual experience with such a monitoring system as this will lead to definitive answers; given the issues involved, the necessary investment seems small.

The electrophoretic screening technique is available as a manual system now and is in the process of being mechanized. Mechanization, however, is best done as part of an ongoing effort. There is therefore every technical reason why screening should be started now and mechanized in stages. Part of the rationale for this is that at each step one must ask: Does the mechanized system yield results identical with those obtained by manual methods? Hence both must be used during the course of developmental studies.

VII. ESTIMATES OF THE MAGNITUDE OF EFFORT REQUIRED

In designing any such monitoring program as this, one must first make a decision concerning the extent of the increase in mutation rates which the program should detect. The magnitude of the effort required then depends on many factors, some known, some to be derived only by experience. The most important of the unknown factors is the spontaneous rate of mutation. Two figures commonly employed in calculations concerning man are 0.5×10^{-5} or 1×10^{-5} *detectable* mutation/locus/generation. Should the spontaneous rate for the type of mutations to be detected by such a program as described prove to be as low as 1×10^{-6}/gene/generation, then a larger sample would be required to detect some arbitrary percentage increase, such as doubling, than if the spontaneous rate were, to pick a value on the high side, 1×10^{-4} mutation/gene/generation.

There would undoubtedly be a wide spectrum of opinion among geneticists regarding the magnitude of the increase which should not be permitted to escape detection. Somewhat arbitrarily, we will select a figure of 50%. Most geneticists would probably agree that it would be unwise to permit an increase as great as a doubling to go undetected—more on intuitive grounds than because the results can be clearly visualized. On the other hand, not only does the detection of increases as small as 10–20% require very large numbers, but such a small increase, while having a definite impact, would to most be temporarily acceptable in light of the returns to society from the processes resulting in the exposure.

Three calculations have appeared concerning the sample sizes which might be involved. Strobel and Vogel in 1958 (*cf.* Vogel, 1970) were the first to compute the sample size necessary to demonstrate a specified increase in the mutation rate. They phrased their problem in terms of mutations resulting in gross phenotypic effects, but the approach is of course equally valid for biochemical mutations. Let us assume, following their approach, that the frequency of mutant phenotypes in the population screened is one in 5000 persons screened (20 proteins, and a spontaneous rate of mutation in the corresponding structural genes of 0.5×10^{-5}). We wish to detect an increase in mutation rate of 50% (in their terminology, a t value of 1.5), with a probability of 95%. Based on formula (7) of Vogel, it can be calculated from these assumptions that if the program design is to monitor successive, equal-sized samples of births, then the two samples need to be of size 274,000. That is to say, if as the result of extraneous influences the mutation rate increased by 50%, under these assumptions this could be demonstrated by two successive samples of 274,000 births.

Crow (1971), in a recent treatment of the same subject, has written:

> Say the normal mutation rate is 10^{-5} per gamete. A population of 3 million births would produce 60 new mutants. The standard deviation is the square root of this, or roughly 8. An increase to 80 would be significant. So, to detect an increase of one-third in the incidence requires the examination of a number of births comparable to the yearly number born in the United States. . . . The efficiency could be increased by searching for mutations simultaneously at a number of genes, but still the undertaking would be very expensive.

We would suggest that one should not proceed with such a monitoring program without a battery of at least 20 indicator proteins. This type of calculation then suggests that, on the assumption of a spontaneous rate of 0.5×10^{-5}, and monitoring of 20 proteins, 250,000 births would yield 25 mutants, and an increase of 13 (50%) would be significant at about the 1% level.

Neel (1971) has presented still a third calculation. The argument proceeded as follows:

> Let us assume that the rate at which spontaneous mutation results in detectable changes in a protein is 0.5×10^{-5}/locus/generation. Let us assume we can monitor for 20 different proteins (some associated, like the α and β chains of hemoglobin), and that we wish to be able, at the 0.05 and 0.01 probability limits, to detect a 50 percent increase in the frequency of mutation. We wish to detect an increase from an estimated 20:100,000 births to 30:100,000 births. If we accept a Type II (β) error of 0.20, this requires at the 0.05 level 365,000 person determinations in each sample or, at the 0.01 level, 592,000 person determinations.

Unfortunately, there was a simple arithmetical error in that calculation;

TABLE 2. Sample Sizes Necessary to Detect an Increase in Certain Biochemical Mutants[a]

Value of α	Value of β		
	0.10	0.20	0.56
0.05	434,000	313,000	135,000
t test	3.18[b]	2.70[b]	1.65[c]
0.01	660,000	509,000	270,000
t test	3.92[b]	4.44[d]	2.33[e]

[a]From an observed frequency of 20:100,000 births to 30:100,000 births, at the indicated probability levels for type I (α) and type II (β) errors. The *t* values are based on one-tailed probability distributions.
[b]$0.005 > P > 0.0005$.
[c]$0.05 > P > 0.01$.
[d]$0.0005 > P$.
[e]$0.01 > P > 0.005$.

the appropriate numbers are 313,000 and 509,000 person-determinations, respectively (see Table 2).

These three calculations are all in reasonable agreement, since Crow's calculation did not include allowance for type II errors. Under these assumptions, to detect a 50% increase in mutation rates two successive samples of approximately 300,000 are necessary, the exact number depending on the requirements one builds into the calculation. Some may feel that it is not sufficient to detect an increase only when it reaches 50%. We suggest that this is a reasonable initial objective, subject to reconsideration with experience.

Let us assume that in this effort to detect a 50% increase we are content to monitor on a 5-year cycle (i.e., contrast the results of successive 5-year periods). We must then collect data on roughly 60,000 births/year. At a birth rate of 20/1000 population, we must accordingly examine all the infants born to populations, however defined, of 3 million persons. A very major effort is required. But is this, as Vogel suggests (1970), "a research project . . . of fantastic dimensions"? Is this approach, as Crow (1971) suggests, "prohibitively expensive"? The answer (assuming, following the preceding section, that the method is sound if not perfect) depends primarily on two factors: (1) *How important it is to have accurate information on this point, and* (2) *how great an effort is involved and what technical developments are in sight to ease the routine burdens of the work.*

VIII. IMPORTANCE OF ACCURATE INFORMATION

With respect to the first question, regarding the real need for accurate data, we are of course confronted with a value judgment. The prospect of a modest increase in mutation rates following increased exposure to ionizing radiation has been sufficient to provoke the formation of a variety of high-level committees. A persistent challenge to the development and application of nuclear energy and power continues to be the possible genetic implications of even small increases over present exposures to radiation. The threat of chemical mutagenesis has now been raised in numerous articles and public hearings.

The present is a time of intense concern for the human animal and what, unwittingly, he may have done to himself and his long-term prospects for survival on this planet. While we do not belong to the camp of extreme alarmists, we do feel that enough legitimate questions have been raised to indicate the possibility of a problem as regards mutation rates. Ours is a society which spent billions to put a few men on the moon, in an undertaking many would feel was characterized more by pride and a concern for public image than a concern for human well-being. Ours is also a society in which a military action which a clear majority of Americans have recognized for some years to be a colossal miscalculation somehow drags on. A very small fraction of the expense of either of these undertakings would ensure this type of study. Biomedical scientists, while talking big concerning the threats to man, are still the victims of a think-small psychology operationally—a psychology the physical scientists have long since overcome. If we can construct great facilities to gain added insight into the particulate nature and alternative states of inanimate matter, surely we can afford equal sums to gain understanding of the particulate nature of man, as expressed through the alternative states provided by mutation.

IX. SOME TECHNOLOGICAL REQUIREMENTS

The second value judgment to be reached in any decision to proceed with such a monitoring program concerns the effort required to obtain numbers of the magnitude we have discussed. Here a critical factor is one's outlook concerning the extent to which automation can improve the efficiency of screening for mutations with qualitative and/or quantitative effects. In this connection, it is important to recognize that approximately half of the qualitative variants of serum or erythrocyte enzymes thus far examined in this respect are also quantitative variants (cf. Harris, 1971).

Techniques for screening for variants in 30 proteins, including 15 of the 17 "best" proteins (20 polypeptides) listed in Table 1, are currently operational in the genetics laboratory at Ann Arbor. On the basis of the experience gained, we estimate that with the present state of the art, one technician can easily screen 1000 bloods for 20 variants a year. With mass production and specialized responsibilities, that number could be doubled. If, following an earlier calculation, our goal was 300,000 person-determinations in some given time interval, this represents 150 technician years, or, on a 5-year cycle, 30 technicians plus the necessary supervisory professionals, perhaps eight in number. In addition, a formidable collecting net must be organized. Let us assume our sample is composed of newborn infants (cord blood) and monitoring is on a 5-year cycle. This requires (allowing for inadequate samples, contaminated samples, etc.) approximately 80,000 specimens a year. Now, there are some 381 hospitals in this country with 2000 or more deliveries a year, for the most part located in urban centers, with a total of 1,165,450 deliveries annually. Thus the requisite sample could be obtained from some 30–40 hospitals, the exact number of hospitals depending on their size and the completeness of collection. The collection of these specimens on a round-the-clock basis would require one person-equivalent in each of these hospitals. Assuming the specimens are sent to a central laboratory (or two), some kind of pickup system must be organized for the participating hospitals, but this presents no particular problem. Finally, provision must be made for obtaining specimens for the necessary family studies when a variant is encountered, as well as for an administrative and clerical staff. We estimate total annual personnel needs (5-year cycle) in the present state of the art to be at least some 120 persons.

While technological developments have little to offer the collection aspects of this operation, they have much to offer the actual processing of blood specimens. The development of supporting media for electrophoresis such as acrylamide gel (discussed in a later section), which is much less difficult to batch process than starch gels, offers promising possibilities for automation in the detection of qualitative variants. Already there exist a variety of mechanized laboratory devices for the determination of the activity of such serum enzymes as alkaline phosphatase, lactic dehydrogenase, glutamic oxalacetic acid transaminase, creatine phosphokinase, aldolase, and isocitric dehydrogenase. Although the range of normal variation in the level of these enzymes is so great that the detection of a genetically determined half or three-quarters normal activity seems unlikely, conceivably a battery of more appropriate enzymes can be developed. Catalase, NADH–diaphorase, and galactose-1-phosphate uridyl transferase activity are examples of enzymes for which the values in heterozygotes for some mutant genes involving those enzymes exhibit relatively little overlap with homozygous normal or homozygous defective individuals (reviewed in Hsia, 1971). Alternatively, as discussed later, new analytical systems may be employed which determine enzyme kinetic constants which are independent of the amount of enzyme present. Certainly, in view of the newness of this field and its speed of advance, it is too early to be pessimistic about the possibility that substantial improvements, including automation of some procedures, will greatly facilitate screening procedures in this field. Any apparent variant with quantitative effects should, of course, also be studied electrophoretically.

To what extent automation can reduce the personnel requirements mentioned earlier is not now clear. A halving of the present requirements for technical personnel seems a very reasonable objective. If the previous experience of technology with the "automation" of a wide variety of procedures is any guide, this will prove to be a conservative estimate.

Lest it be thought that large-scale repetitive biochemical or immunological testing is new and formidable, note that in the Netherlands over a million potato plants have been tested yearly for potato viruses X and S by agglutination or microprecipitin techniques. These tests are all done in a 3- to 4-week period with substations performing 21,000 tests/day during this interval and consuming 160 liters of immune serum each year (Ball, 1964).

If we admit that *mass* genetic monitoring or screening will become an important field in its own right, as will be discussed in detail in a later section, then there may be a logic which is natural to the field. It is instructive to consider the history of screening in clinical chemistry. Briefly, a series of tests have been developed, usually involving major constitutents of blood or urine, which could be analyzed by using available samples, reagents, procedures, and instruments. These have come into wide use as clinical tests for quantitative health screening. There is no guarantee that they are by any means the most important tests that could be done from either a genetic or a medical viewpoint. The logical approach, in contrast, would appear to be to ask how many different substances are present in the screening sample, then to find out how many of these are associated with genetic changes and human disease. We would thus increase the probability of finding genetic variants in a smaller population screen and increase our knowledge concerning biochemical aberrations in man leading to disease states at the same time.

Whether one is approaching mass genetic screening or mass health screening, the problem of finding out what substances *should* be included in a screen is quite different from the question of screening for one or a specified list of substances. In the former, one wishes to quantitate or characterize a large number of different substances; in the latter, after one has selected a few key substances to measure, he needs simple specific methods applicable to a very large number of samples. We define the former as "scanning" methods, while the latter are true "screening" methods. There is always the possibility that new technology may bridge the gap between the two and allow a very large number of substances to be determined on a very large number of samples.

One should extend this discussion to include scanning a large number of compounds to determine qualitative changes that would indicate those constitutents in the sample that would be suitable for detection of genetic variants and ultimately as markers for determining mutation rate due to exposure to harmful chemicals or ionizing radiation in the environment. Specific mechanized or automated screening methods could then be established to monitor qualitative changes in these substances. The automated techniques should emphasize the need for quantitative as well as qualitative data from the chosen automated procedures to provide both information for health care and information concerning human genetics. In the present context, we need a suitable compromise between scanning and screening methods. The effort involved in obtaining and preparing a sample for analysis is such that principles of efficiency call for as many simple determinations yielding clear results as can be mustered.

X. ASSESSMENT OF TECHNOLOGICAL POSSIBILITIES (USING ELECTROPHORESIS)

Thus far, the discussion of the technical problems in monitoring for mutation has largely been addressed to the present state of the art. However, important technical developments which would have implications for such a program are imminent. Some of these are exemplified by techniques under consideration in connection with "health screening."

In this section, we will explore technical requirements and possibilities, always seeking out those problems which would give the greatest difficulty in practice. If solutions to these can be devised, then it appears reasonable to initiate a full-scale effort. Hence, the reader must not expect the description of finished systems in every instance; rather, we will examine a series of specific problems to see if plausible solutions exist.

A. Sample Collection and Preparation

Most automated instruments available for screening procedures or high-volume analyses of samples involving a complex biological matrix, such as whole blood, serum, or plasma, are designed after the fact when it comes to sample collection and sample preparation. They assume a readily available hemolysate, a serum sample separated from its clot, or plasma separated from red blood cells; although instrumentation built on these assumptions represents a definite saving of labor plus generally improved precision in the analytical results due to decreased errors caused by fatigue, the preparation of the sample for analysis, while maintaining sample and test identification on each sample, still represents a bottleneck in the automation of screening methods. Some effort has been devoted to the development of a system which will allow the dynamic introduction of whole blood into the analytical system and permit several different tests to be performed simultaneously on the separated plasma without any further sample handling (Scott and Mailen, 1972). Likewise, work has been initiated on the development of an automated system for the separation, washing, and hemolysis of red blood cells (Fig. 1) (Anderson and Willis, 1973); this should be a welcome addition to sample handling for automated screening procedures. Much attention is still required in this area of automation before adequate systems will be developed that will allow screening procedures to proceed directly from the collected sample to the analysis which includes an integrated sample preparation step.

Centrifugation of samples with manual introduction and retrieval of the individual sample or groups of samples is still the mode of operation in most biochemical and clinical laboratories, and there exist a variety of automated dispensers for the introduction of these samples into continuous-flow, discrete, and parallel automated analyzers. However, the tedium of application of a quantitative amount of sample to a medium for electrophoresis has as yet no practical solution suitable for mass use. An

FIGURE 1. Prototype system for washing a series of different erythrocyte samples in parallel, hemolyzing them, and then recovering them for electrophoretic analysis.

automated thin layer chromatography apparatus now exists that includes automated sample identification and application, and automation has been perfected for the introduction of samples into gas–liquid chromatographs as well as onto ion exchange chromatography columns. In concluding this discussion of automated sample introduction, some note should be made of attempts to automate liquid–liquid solvent extraction to separate sample components prior to analysis (Wallace, 1967).

More careful evaluation must be made in sample collection and storage regarding the questions of which anticoagulants to use, whether both plasma and serum are required, and whether freezing or refrigeration in a nonfrozen state is a better method. Methods for the storage of the sample until used to maintain the best possible integrity of sample components have been considered in some detail; for reviews, see Mazur (1970), Chilson et al. (1966), Rowe et al. (1968), and Rowe (1971).

B. Electrophoretic Separation

The objective is to determine whether the electrophoretic mobility of a given protein is altered in any one of a large number of samples. The underlying operational assumption is that some identifying aspect of the protein in question is not altered by the mutation. Extreme examples of instances where a mutation would nullify the detection process would be a mutant hemoglobin which did not bind heme (this could only be a small percentage of the total hemoglobin or it would be lethal) or a mutant enzyme having a nonfunctional or subfunctional active site. Many inborn errors of metabolism fall in the latter class.

The methods described in this section involve the measurement of electrophoretic mobility using (except for hemoglobin) some method of color production to locate individual proteins. The first question to be asked is whether sufficiently high-resolution separations and identification by position alone can be achieved to allow histochemical identification to be dispensed with. Using high-resolution acrylamide gel electrophoresis, a large number of serum proteins may be seen as separate bands, with estimates of the total number resolvable varying from 30 to 60 (Caton, 1972; Wright et al., 1971). Quantitative staining methods and automatic scanning devices have been developed which allow reasonably precise quantitation of major bands (Caton et al., 1972). The methods are equally adaptable to extracts of cells and have been widely employed to study red cell hemolysates (Strickland, 1970). For a few major components including serum albumin, simple measurements on stained gels suffice. For the remainder, the normally occurring different transferrins, fetal proteins such as α- and β-fetoprotein (Abelev et al., 1963; Vasileiskii,

1965), and proteins of placental origin (Hoffman *et al.*, 1970) found in serum during pregnancy are a source of confusion, and some method of specific band identification is required.

1. Electrophoresis Without Specific Detection Reactions

Since monitoring for mutants by electrophoretic techniques may ultimately be performed in conjunction with screening newborn infants for genetic disorders, and since *some* variants may be detected by electrophoresis, we have asked where the major difficulties lie in adapting the method to mass use. We conclude initially that some supporting material such as starch or acrylamide will be used for all electrophoretic separations. The support may be in small tubes, cast as slabs, or applied to long strips of film; each method has its own advantages. Since the highest-resolution separations appear to employ gradient gels, we have asked whether large numbers of such gels could be made simultaneously. A centrifugal gel-casting rotor (Fig. 2) was constructed which holds 500 glass disc-electrophoresis tubes (Caton and Anderson, 1973). During rotation, an acrylamide gradient is pumped into the rotor edge through a center seal, and all tubes are filled simultaneously with identical gradients. Rotation is continued until polymerization is complete, at which time rotation is stopped and the tubes are removed and stored under buffer until ready for use. The patterns ob-

FIGURE 2. Centrifugal gel-casting rotor for producing 500 gradient gels at one time.

tained with these tubes with serum are reproducible. This rotor was constructed in an orienting feasibility study which aims at the development of a centrifugal system for performing all of the steps involved in gel casting, sample introduction, electrophoresis, fixing, staining, destaining, and possibly scanning as one continuing operation with most steps occurring in a centrifugal field. Other concepts for casting continuous films and large numbers of gel blocks have also been explored, and it is quite evident that a number of alternate approaches are now available which need to be reduced to practice and compared.

2. *Adaptation of Histochemical Enzyme Reactions to Electrophoretically Separated Proteins*

A wide variety of reactions applicable to starch or acrylamide gels have been listed or referenced in a previous section, and these will not be discussed in detail here. The problems associated with using these reactions with large numbers of samples are the following:

1. Either the reactants must be applied in parallel along a flat electrophoresis gel, or the gel must be cut up to give a number of presumably identical gels which are treated separately.
2. Some reactions require the addition of a single reagent, others require the sequential application of two or more.
3. Reaction times differ for different activity stains.
4. Reagents vary in stability; therefore, some must be prepared at frequent intervals.

An alternative to be given serious consideration is that the electrophoresis of one sample is done on a large slab or film, and all, or most, reagents are applied to this one slab. Then sample identification is easier than if each block is sliced to produce many, each of which is run through a separate process. If possible, therefore, it would appear advantageous to attempt to do all reactions on one slab and keep all information confined to one pattern which may be more readily intercompared.

To explore methods for doing this, we have infiltrated the supporting medium (agar, starch, acrylamide) into folded or rolled paper or other fibrous support. The electrophoresis is done in the compact form, then the fibrous support is unrolled and treated with histochemical reagents applied in parallel paper strips. The strips may be removed at different times or replaced by others containing different reagents. Much further work remains to be done to extend these exploratory studies to the production of working prototypes. However, no insoluble problems appear to have been uncovered thus far.

C. Detection Using Immunological Methods

1. Immunoprecipitation

Immunoprecipitation methods may be employed in conjunction with electrophoretic separations in at least two ways: In the first, specific antibodies or mixtures of them are used to form precipitation arcs with the individual electrophoretically separated proteins. As the method is ordinarily employed in immunoelectrophoresis, small differences in electrophoretic mobility are difficult to detect. However, modifications have been developed which make the method more precise. For example, chimpanzee transferrin, haptoglobin, and α_2-macroglobulin have mobilities different from their human homologues. When human and chimpanzee sera are run in parallel, and antiserum is diffused in from both sides and allowed to form complete ovals around the antigens, these mobility differences are readily observed (Fig. 3), as shown by Williams and Wemyss (1961). Methods for increasing the resolution of this system are now under development in the MAN Program laboratories at Oak Ridge and may solve the problem.

The pattern seen in Fig. 3 is complex and may be much simplified by including only antibodies against a selected subset of proteins. Alternatively, one or more different fluorescent labels may be attached to selected antibodies, allowing their unambiguous identification.

The second general method is to use antibodies to precipitate enzymes which are then identified by the techniques of enzyme histochemistry. A

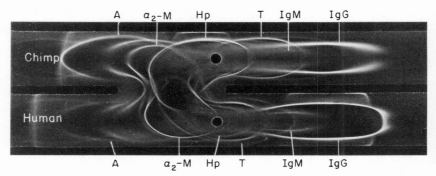

FIGURE 3. Method for demonstrating mobility differences between individual proteins in two mixtures. Illustrated by interaction pattern of chimpanzee and human serum samples in adjacent wells. A, albumin; α_2-M, α_2-macroglobulin; Hp, haptoglobin; T, transferrin; IgM, macroimmunoglobulin; IgG, 7S immunoglobulin. Chimp transferrin (T) has lower mobility than human transferrin; chimp haptoglobin (Hp) has higher mobility and is greatly elevated due to recent infection; chimp α_2-macroglobulin has much higher mobility than the human homologue. (From Williams and Wemyss, 1961).

TABLE 3. Enzymes Identified by the Catalytic Properties of the Enzyme–Antibody Precipitates Obtained in Gel Diffusion Media [a]

EC No. [b]	Enzyme	Substrate
1.1.1.27	Lactate dehydrogenase	L-Lactate
1.1.2.49	Glucose-6-phosphate dehydrogenase	Glucose-6-phosphate
1.2.3.2.	Xanthine oxidase	Xanthine or xanthosine
1.4.1.1	Alanine dehydrogenase	L-Alanine
1.4.1.2	Glutamate dehydrogenase	L-Glutamate
1.10.3.2	Ceruloplasmin	p-Phenylenediamine
1.11.1.6	Catalase	H_2O_2
1.11.1.7	Peroxidase	Benzidine
2.7.7.8	Polynucleotide phosphorylase	Adenosine-5-diphosphate
2.7.7.16	Ribonuclease	Cyclic cytidylic acid
3.1.1	Carboxylic ester hydrolases	2-Naphthyl acetate
3.1.1	Carboxylic ester hydrolases	1-Naphthyl acetate
3.1.1	Carboxylic ester hydrolases	Indoxyl acetate
3.1.1.3	Lipase	2-Naphthyl nonanoate
3.1.1.7	Acetylcholinesterase	Acetylthiocholine
3.1.1.8	Cholinesterase	Butyrylthiocholine
3.1.1.8	Cholinesterase	Carbonaphthoxycholine
3.1.3.1	Alakaline phosphatase	Naphthol AS-MX phosphoric acid
3.1.3.2	Acid phosphatase	β-Glycerophosphate
3.1.3.2	Acid phosphatase	Naphthol AS-BI phosphoric acid
3.1.3.9	Glucose-6-phosphatase	Glucose-6-phosphate
3.2.1.1	α-Amylase	Soluble starch
3.2.1.2	β-Amylase	β-Amylose
3.1.2.23	β-Galactosidase	6-Bromo-2-naphthyl-β-D-glactoside
3.2.1.31	β-Glucuronidase	6-Bromo-2-naphthyl-β-D-glucopyranoside
3.2.1.30	β-Acetylglucosaminidase	Naphthol-AS-LC-acetyl-β-D-glucosaminide
3.4.1.1	Leucine aminopeptidase	Leucyl-4-methoxy-2-naphthylamide
3.4.2.1	Carboxypeptidase A	Carbonaphthoxy-DL-phenylalanine
3.4.2.2	Carboxypeptidase B	Hippuryl-L-arginine
3.4.4.1	Pepsin	Bovine serum albumin
3.4.4.4.	Trypsin	Benzoyl-D,L-arginine-2-naphthylamide

[a] From Williams and Chase (1971).
[b] Code number assigned by the Enzyme Commission (EC) of the International Union of Biochemistry.

TABLE 3.   Cont'd

EC No.	Enzyme	Substrate
3.4.4.4	Trypsin	p-Tosyl-L-arginine methyl ester
3.4.4.5	Chymotrypsin	Acetyl-D, L-phenylalanine-2-naphthyl ester
3.4.4.5	Chymotrypsin	Acetyl-L-tyrosine ethyl ester
4.1.2.7	Aldolase	D-Fructose diphosphate
4.2.1.1	Carbonic anhydrase	H_2CO_3

large number of enzymes are not inactivated by combination with specific precipitating antibodies, and a list of enzymes which may be identified in precipitates is given in Table 3.

Methods based on immunoprecipitation assume that the antigenic site is not altered by mutations, and those based on enzyme histochemistry of precipitates assume that neither the antigenic site(s) nor the enzyme active site is substantially altered.

2.   *Immunosubtraction*

One problem with high-resolution electrophoresis is that a very large number of proteins are distinguished, and identification of many minor ones, especially after minor changes in technique have been made, is difficult. To help solve this problem, immunosubtractive methods have been developed. Antibodies against one serum protein are immobilized on a solid support in the sample zone. As a result, one specific protein is selectively removed from the sample *before* electrophoresis starts (Fig. 4). Comparison with an unsubtracted serum shows which protein is missing, thus providing positive identification of one band.

FIGURE 4.   Immunosubtraction in gel electrophoresis. Antibodies against transferrin were used to selectively remove it. Control serum is at the top. Increasing quantities of immobilized antibody were used in lower three patterns.

The method may be reversed and *all* but one or a select few removed. This technique was originally employed by Abelev *et al.* (1963) to subtract all but α-fetoprotein from rat and human fetal serum. Preparation of antisera is the major problem. In the Oak Ridge laboratory, this has been solved by passing anti–whole human serum through an affinity column containing one or a few human serum proteins to be detected. Antibodies against these human proteins are thus quantitatively removed. The eluted antibodies in turn may be used to subtract the serum proteins of interest from whole serum to prepare an antigen mixture used for further immunizations. In this manner, large amounts of anti-all-but-one, or anti-all-but-a-select-few antisera may be prepared.

When IgG isolated from these antisera is immobilized on a solid support or cross-linked to form particles which will not penetrate the gel, it will subtract out all but the one or more proteins of interest, which are then studied with maximum resolution by acrylamide gel electrophoresis.

Note a singular advantage of this method: Nothing is required to interact with the protein being examined except a nonspecific stain. The protein may have lost enzymatic activity, and its antigenic sites, and still be detected. Immunochemical methods also have high specificity and allow many different proteins to be identified by one type of reagent, IgG, which is very stable and can be easily prepared in large quantities.

There is a conceptual contradiction in methods which employ diffusion-dependent reactions after electrophoresis. The objective in electrophoresis is to minimize diffusion and obtain as high resolution as possible. Immunoprecipitation, as ordinarily done, and enzyme histochemistry are diffusion dependent and require that time be allowed for reactants to diffuse into the gel and for reactions to occur. Immunosubtraction minimizes these resolution-blurring processes and only requires time at the end of electrophoresis for hydrogen ions or other fixing substances to diffuse into the gel.

XI. METHODS BASED ON ACTIVITY MEASUREMENTS

While methods based on electrophoresis have first priority, a number of other methods, requiring very small samples and capable of adaptation to the recently developed computer-interfaced fast analyzers (Anderson, 1969*b*, 1970), could be easily added to the monitoring program.

A. Enzyme Variant Scanning

Automated enzyme activity assays have been a valuable aid to clinical diagnosis (Coodley, 1970) but have been used only to a limited extent in the detection of mutants. Part of the problem is that there are a variety of factors contributing to the level of serum enzymes (Posen, 1970) that are not of direct genetic relationship.

The use of enzyme activity assays in the presence and absence of inhibitors has some merit and can be automated. Cholinesterase variants have been detected by Kalow and Genest (1957) and Harris and Whittaker (1961) using dibucaine and fluoride inhibition, respectively; they have provided a means of obtaining both qualitative information (type of variant) and quantitative information (activity of cholinesterase in the sample) about the enzyme in serum.

The kinetic parameters of an enzyme—the Michaelis constant, K_m, and the maximal velocity, V_{max}—should provide both qualitative and quantitative information concerning variation of an enzyme due to the change in the enzyme's kinetic properties brought about by conformational changes in the tertiary structure of the enzyme induced by substitutions, deletions, or addition of amino acids. Since the K_m of an enzyme is a qualitative property of the enzyme and does not vary with enzyme concentration, it could be used as a qualitative monitor for genetic variation. Both the quantitative information (V_{max}) and the qualitative information (K_m) potentially can be obtained rapidly in an automated mode, as will be discussed in the following section on isozymes, making this an interesting complementary approach to electrophoretic scanning. Motulsky and Yoshida (1969) have shown that in the case of glucose-6 phosphate dehydrogenase in red blood cells, K_m variants are directly related to genetic variants, while Davies *et al.* (1960) have shown the same phenomenon for cholinesterase variants.

B. Isozyme Scanning

As previously discussed, those isozymes which have frequently occurring variants make it difficult to identify rare variants if the rare variant has an electrophoretic mobility similar to the mobility of one of the polymorphic forms. However, the kinetic parameters of an enzyme, in this case isozymes, offer a potential means of scanning for these rate variants as well as being useful for diagnostic health screening. The determination of the kinetic parameters (K_m, V_{max}, etc.) of isozymes using steady-state kinetics presents itself as a special case, with mathematical resolutions existing for two isozymes present in a sample (Cleland, 1970). Although the relative concentrations of the ioszymes may vary from sample to sample, which will be reflected by changes in V_{max} for different samples, the K_m of each of the two isozymes is an individual property of the respective isozyme and does not vary with concentration. Thus K_m has potential use as a qualitative scanning method for isozyme variants.

Literature abounds with kinetic, spectrophotometric, and fluorimetric assays for a variety of enzymes; for reviews, see Bergmeyer (1963) and Roth (1971). The kinetic scanning of isozymes for variants can be automated in a variety of ways, such as using continuous flow or using parallel analysis with a multicuvette spectrophotometer in conjunction with a small on-line computer. Such instrumentation, the GeMSAEC Fast Analyzer, has been developed at ORNL (Anderson 1969a, b, 1970). Consisting of either a 15– or 42– cuvette spectrophotometer interfaced to a computer, it provides rapid addition and mixing of all samples and reagents in parallel using centrifugal force. With this type of instrumentation, the various substrate dilutions for the determination of the kinetic parameters of an enzyme or isozyme can be run in one experiment. For example, an oscilloscope plot of absorbance *vs.* time of reaction for placental alkaline phosphatase at several different concentrations of *P*-nitrophenyl phosphate is shown in Fig. 5. One type of reciprocal plot, substrate/velocity *vs.* substrate concentration (S/V *vs.* S), automatically plotted from the initial velocity data of Fig. 5, is shown in Fig. 6. Plots of S/V *vs.* S, $1/V$ *vs.* $1/S$, or both are available. Automated data reduction procedures patterned after those of Wilkinson (1961) are used to provide a best-fit value for K_m and V_{max} in conjunction with the standard error of

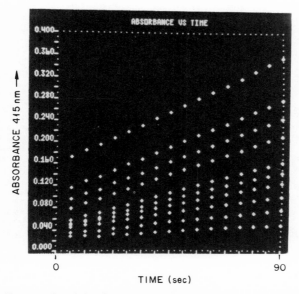

FIGURE 5. Oscilloscope plot of absorbance *vs.* time for reaction of placental alkaline phosphatase with different concentrations of *p*-nitrophenyl phosphate using ORNL GeMSAEC system.

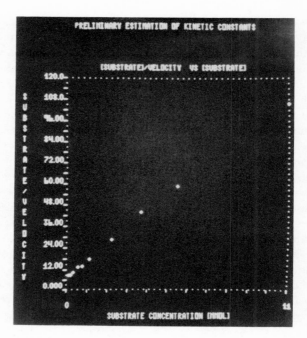

FIGURE 6. Plot of substrate/velocity *vs.* substrate concentration (*S/V vs. S*) for data from Fig. 8.

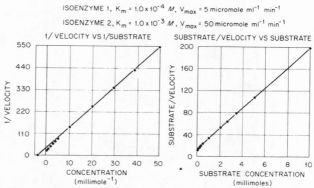

FIGURE 7. Preliminary estimate of kinetic constants (see text). The effect on linearity of reciprocal plots due to the presence of isozymes.

estimation for both K_m and V_{max}. Total analysis time, from start of the instrument to final printout of the data and publishable photograph of reciprocal plots, is of the order of 10 min. The effect on the linearity of two types of reciprocal plots due to the presence of two isozymes, one with

TABLE 4. Apparent K_m and V_{max} for a System Composed of Two Isozymes as a Function of Isozyme Concentration

Isozyme 1 activity[a] (μmole ml^{-1} min^{-1})	Isozyme 2 activity[b] (μmole ml^{-1} min^{-1})	Apparent V_{max} (μmole ml^{-1} min^{-1})	Apparent K_m (mmole)
100	50	128	0.136
50	50	80	0.176
25	50	56	0.246
5	50	52	0.707

[a] $K_m = 0.01$ mmole.
[b] $K_m = 1.0$ mmole.

a K_m of 0.1 mmole and an activity of 5 μmole ml^{-1} min^{-1} and the other with a K_m of 1.0 mmole and an activity of 50 μmole ml^{-1} min^{-1}, is shown in Fig. 7. The curvature, although present in these plots, is not grossly apparent. Table 4 demonstrates the variation in the apparent K_m of such a system without an attempt to separate the two K_m values of the isozymes by mathematical means. However, a relatively simple program exists to automatically separate the two K_m's, thus preventing an apparent concentration variation of the isozyme K_m's and preventing the reporting of the pseudo-K_m variants when they do not indeed exist.

Whether or not this will become a scanning procedure for enzyme variants rests with its experimental success or failure. If it should succeed, the potential of handling a large number of samples per day and rapid output of the data exists and should provide some attraction for its use.

C. Enzyme Group Evaluation

Methods for evaluating the status of enzymes in the glycolytic cycle in red cell hemolysates have been described using simple fluorescence spot tests which allow detection of enzyme defects in a group of enzymes, with determination of the specific defect deferred until later and applied only to those individuals found to be abnormal by the screening methods employed (Yunis, 1973).

While this does not constitute an exhaustive survey of the status of scanning for disease-associated and/or genetic defects, it is sufficient to show that some of the tools are at hand to allow the evaluation of many different possible indicator substances for inclusion in screening tests.

XII. LOW MOLECULAR WEIGHT SUBSTANCES

A. Scanning for Screen Evaluation

Every metabolite and most excretion products are potential indicators of genetic defects. Once a monitoring system based on cord blood has been established, a relatively small incremental expenditure could ensure the collection of a urine specimen from each child being monitored. The Body Fluids Analysis Program at Oak Ridge has been concerned with the analysis of human urine and serum using high-resolution chromatography. Untraviolet-absorbing compounds have been of major interest, and over 90 such compounds have been identified. Metabolic (presumably genetic) defects studied thus far are listed in Table 5; this list is being rapidly enlarged under a collaborative program involving ORNL, the NIH Clinical Center, Duke Medical School, and Johns Hopkins Medical School. A high-resolution chromatogram of a human urine is shown in Fig. 8.

High-resolution chromatography of carbohydrates in human urine on borate anion exchange columns has revealed that a remarkably large number of sugars are present, as is illustrated in the lower chromatogram in Fig. 9. Galactosemia variants are shown in the upper two chromatograms of the same figure and demonstrate both qualitative and quantitative differences from the normal composite urine. In addition to the metabolic defects expected (e. g., galactosemia, lactase deficiency), further exploratory studies with this type of system may be expected to turn up new inborn errors of metabolism as well as other information concerning human and mammalian biochemistry. For example, the presence of sucrose in human urine as demonstrated by high-resolution carbohydrate chromatograms emphasizes that certain aspects of current knowledge about human biochemistry may be deficient, partly due to nonspecific measuring techniques. Sucrose excretion in the urine, apparently of endogeneous origin and previously associated with disease of the pancreas, has been reported by Rosenfeld et al. (1965). There is some indication that sucrose in normal urine is not entirely of dietary origin (Gorodezki et al., 1971) and that its location of synthesis is in the kidneys. Studies

TABLE 5.   Inborn Errors of Metabolism Detectable by Screening Urine

Alcaptonuria	Maple syrup urine disease
Diabetes mellitus	Orotic aciduria
Galactosemia	Phenylketonuria
Hyperoxaluria	Renal glycosuria
Hypertryptophanemia	Tyrosinosis

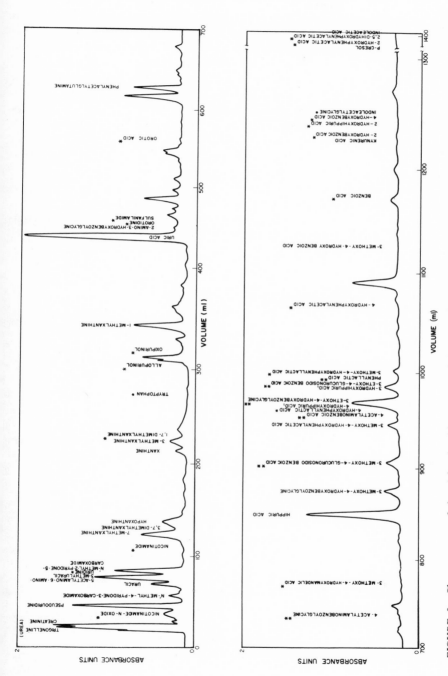

FIGURE 8. Chromatogram of ultraviolet-absorbing compounds in normal human urine using ORNL UV ion exchange chromatograph.

performed by Young *et al.* (1971) on dietary effects on urinary compounds show a decrease but not an elimination of urinary sucrose with special diets. Thus the point of this discussion is that sucrose is a nonreducing sugar and cannot be detected clinically using standard nonspecific reducing methods; its presence as a component of human urine could not be readily determined previously.

In addition to endogenous biochemicals and exogenous dietary biochemicals observed in the urine, many drugs and drug breakdown products are observed in these chromatograms, making them useful in pharmacological and toxicological studies (Burtis *et al.*, 1970). Ninhydrin-positive compounds, such as amino acids and biogenic amines, have been identified in the urine; this has led to the development of sensitive fluorimetric methods for detecting tyrosine and tryptophan metabolites in urine, plasma, and platelets using coupled anion–cation exchange chromatography (Chilcote and Mrochek, 1972).

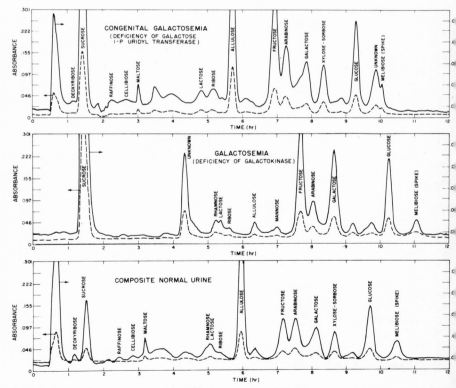

FIGURE 9. Chromatogram of carbohydrates in normal human urine (bottom) and in urine of two patients with galactosemia using carbohydrate analyzer developed at Oak Ridge.

A large amount of qualitative and quantitative data is generated per chromatographic run. The utility of the high-pressure anion and cation exchange chromatographic systems in providing information concerning a large number of different compounds in urine has been greatly enhanced by the addition of a small special-purpose computer that is capable of reducing a chromatographic envelope to Gaussian peaks from the analog signal of the chromatograph's detector and transforming it into digital information such as elution time and amount of each detected compound (Scott *et al.*, 1969; Chilcote and Scott, 1971; Chilcote and Mrochek, 1971).

In addition to substances separated initially by liquid chromatography, volatile substances or volatilized derivatives may be studied directly using gas–liquid chromatography. More recently, mass spectrometers have been coupled to the gas–liquid chromatograph to improve the ability to identify unknown components in the sample (Jellum *et al.*, 1971). Gas–liquid chromatographs coupled with intermediate- and high-resolution mass spectrometers and interfaced to small dedicated computer systems, such as the Digital Electronics Corporation (DEC) PDP-12 computer, are becoming more common. The development of disc-stored libraries of mass spectra data of a large number of low molecular weight compounds, in conjunction with availability of rapid search and comparison routines, promises to make this an indispensable scanning procedure for the identification of low molecular weight compounds in physiological fluids.

B. Automation of Screening Methods for Low Molecular Weight Substances

The automation of chemical substrate analysis has undergone considerable change in the past several years. Continuous-flow and discrete sample analyzers are undergoing constant improvement. Now 15–20 substances may be determined on samples of 1 ml or less. These, however, are still an order of magnitude away from present requirements. The ultimate question is whether the large number of determinations will be based on a large battery of different, presumably specific tests or on sensitive separation methods followed by simple monitoring for a common group or property. The time difference at present is approximately 3 orders of magnitude. Future research may reduce this difference considerably. As an example, the separation of nucleotides has been reduced to minutes using liquid chromatography with pellicular resin as a support (Burtis *et al.*, 1970). The GeMSAEC Fast Analyzer, as previously mentioned, allows samples in the 1–50 µl range to be analyzed in parallel with direct computer interfacing and data reduction. As many as 40 samples may be analyzed in parallel, and for some reactions the total analysis time for the 40 samples may be as little as 15 sec.

Thin layer chromatography, as was previously mentioned, has been automated, with the main emphasis on drug detection. However, it seems reasonable to expect this system to be able to handle carbohydrate and amino acid detection as well, and this would have a practical impact on mass screening for genetic variants.

Automation of gas–liquid chromatography, including sample extraction, has recently been introduced; it also has had as its main emphasis the screening for drugs of abuse. However, the system appears to be flexible enough to perhaps allow automated analysis of other urinary or serum and plasma components, such as steroids, that would have some bearing on genetic variation.

XIII. CONFIRMATORY ANALYSES, GENETIC FOLLOW-UP, AND DATA REDUCTION

When a mutant protein is found, it should be rerun under very carefully controlled conditions to confirm that an altered substance is indeed present. Following confirmation, it is important to determine, when sufficient protein can be obtained, what amino acid substitutions have occurred, since this provides detailed insights into the precise effects of a mutagen. To do this, it is necessary to isolate the protein in pure form— this is most easily done, once the reagents are available, with an immunosubtractive affinity column. The protein is then digested with trypsin and fingerprinted to see if an altered peptide is present. The altered peptide may then be sequenced.

As previously discussed, additional samples from parents, siblings, and offspring will be sought. This involves feedback of information to the donor and some form of genetic counseling. The individuals involved need reassurance where there are no pathological implications and professional advice and help where the opposite is true. This means some direct interfacing with the health care system so that all data obtained may be most effectively used.

Provided that the problem of an identifying number (usually the Social Security Number) is solved, there appear to be no insurmountable problems in record keeping and data reduction. There are many knotty problems to be solved if the data are to find their way into patients' records, whether computerized or not. However, considerable efforts to solve the computerized patient record problem are being made by the Health Services and Mental Health Administration and by several computer manufacturers. It does not appear that data reduction will be an insoluble problem.

XIV. GENETIC MONITORING AS A NATIONAL PROBLEM

It is not clear that public concern over the prospects of an increased mutation rate will by itself result in support of the necessary studies. However, a program such as this is readily combined with one monitoring for a variety of inborn errors of metabolism. Already a variety of such programs have been justified in their own right, such as those in Massachusetts (Shih *et al.*, 1971; Levy *et al.*, 1972), in New York (Scheinberg, 1971; Guthrie and Murphey, 1971), in Quebec (Scriver, 1972), in Austria (Thalhammer *et al.*, 1970), and in New Zealand (Houston and Veale, 1971; Veale *et al.*, 1971). Although for maximum efficiency these programs utilize a blood sample drawn at age 5 days (just prior to discharge of mother and child from hospital), there are already six diseases (admittedly very rare) which can be diagnosed on the basis of a sample of cord blood.

When we consider the possibility of mass multiparameter screening, probably supported by health maintenance organizations (HMO's), in which many analyses are done using very sophisticated technology, where the possibility exists for detecting large numbers of different genetic and nongenetic diseases in addition to mutations, and where methods used for genetic screening are the same ones used for other screening techniques, then we conclude that *monitoring for mutation rates as part of mass health maintenance screening is inevitable.*

It would be a mistake to foster the expectation of mass screening if the technology is neither available nor within sight. The type of technology required depends on whether the screening is to be done locally, regionally, or in only one or a very few centers; this in turn depends in part on the benefits gained from slow *vs.* fast analysis and data return. If some genetic defects may be seen which require prompt treatment, then it is important to make the analytical information available rapidly. In order to plan and develop a screening system, a variety of problems existing at the national, regional, and local levels and extending to the education of the individual physicians and patients must be solved to be certain that no misunderstandings of either intent or results occur. As always, however, the overriding factor will be cost.

A. Cost Benefit Considerations—Does the Potential Total Benefit Exceed the Total Cost?

We distinguish three types of benefits: (1) direct benefit to the individual (sample donor) in the form of information to the attending physician and improved counseling or therapy, (2) benefits to society in the form of reduced disease load, cost, and unhappiness, either as a result of benefit 1 or in consequence of the regulation of exposure to mutagenic compounds, and (3) benefits to the scientific, medical, and decision-making communities in the form of precise data on mutation frequency, genetic disease incidence and identification, incidental epidemiological data on nongenetic diseases, basic information on the variability of the human organism, new information on the natural history of human disease, and new data on human biochemistry resulting from the development of higher-resolution analytical systems. Decision-making must be based on evaluation of all these benefits.

Does the benefit from any one of these exceed the cost? Detailed cost/benefit analyses in this area are not possible now and may never be, because they cannot objectively include evaluations of either human happiness *vs.* pain and suffering or the effect of human expectations, since the health care system is "anxiety driven."

The first general conclusion from reviews of mass biochemical screening is that the tests employed now are not sufficiently sophisticated but that new improved tests of greater specificity must be developed together wtih means for making these tests cheaply and reliably. So-called multiphasic health testing, while too costly for general application *now* in the view of many, will come not only because it will be beneficial but also because of rising public expectations. The present state of affairs is best exemplified by the personal observation that individuals who can find little objective justification for mass searches for rare diseases are usually interested in having these tests run on themselves. Each of us would like to be an exception to the rule that cost affects health care delivery quantitatively and qualitatively.

Second, the patient or subject should be his own control. The work of Williams and others (Williams *et al.*, 1970) shows that many values obtained in the clinical chemistry laboratory which may fall in or near the abnormal range for a large population are normal for the individual tested in the sense that in him they are not associated with disease. A shift *within* the normal range of an individual, however, may have serious implications for an individual who has a long history of being at one end of the normal range. If one accepts the thesis that *change* may be as important as an absolute value, then one desires as complete a biochemical history of an individual as possible—beginning at birth. The analysis of cord blood and of early urine samples may therefore be the logical start of a medical history.

B. Benefits to Patients

1. Direct Benefits

A variety of studies have been done to determine what individual benefit results from mass screening; as noted, these suggest that we need new and more specific tests. The direct patient benefit arises from the detection of *treatable* inborn errors of metabolism, from genetic counseling, from the diagnosis of unsuspected illness, from the provision of baseline data against which subsequent tests can be compared, and from eventual acquisition and storage of information which may be of use in emergencies. Thus if very simple, low-cost methods for tissue typing should become available and be included in the screen, the problem of searching for tissue donors (or of being a donor) could be greatly simplified. Since maternal blood and urine could also be included in the screen, all the above apply equally well to the mother.

It must be emphasized that the development of a biochemical-genetic screen which is worth the cost is an ongoing evolutionary process which must start, on a modest level, with what is available. It assumes that the large expenditures for basic biochemical or molecular studies *will* provide new insight into disease, resulting in new diagnostic tests of greater precision (i.e., that research is worth the cost).

We do not here review in detail the value of individual tests or groups of tests. Rather, we stress that test evaluation must be a continuing large-scale effort and that such evaluation can only do the patient good. The technical conclusion is that since the number of tests done will probably increase rapidly, the amount required for any given test must decrease proportionately (i.e., the pressure toward microminiaturization will increase).

If the individual is to achieve maximum benefit, the results must be available to the attending physician rapidly and in an intelligible form. The data must also be available at any later date to any individual-designated physician anywhere.

2. Disadvantages to the Individual

A comprehensive biochemical survey of newborns and of mothers will turn up a large number of minor anomalies which neither suggest nor justify any known therapy. In many instances, the early history of a disease is not sufficiently well known to allow its identification in initial stages. Thus both anomalous qualitative values and true genetic abnormalities may be discovered which can only contribute to an individual's anxiety. For this reason, *data* must be distinguished from *information*, and both must reach the patient, if at all, through the attending physician. We define information as data referenced to some previous work which allows a conclusion to be stated, preferably as a probability, as, for example, "This patient has a 70% chance of having a kidney disorder on the basis of tests done." The evaluation of the patient-oriented use of data and information is an essential part of any screening evaluation study.

C. Benefits to Society

1. Reduced Disease Load

A basic assumption is that early disease detection results in a reduced total disease load because disease is treated earlier and more effectively. Unfortunately, for many diseases no effective treatments are known. In place of a reduced disease load, precisely the opposite may occur, namely, a sharp rise in total identified disease merely because of increased rate of diagnosis. However, evaluation of early diagnosis cannot be made on a short-term basis, since an increase in reported and treated disease now may result in a decline later. A reduced disease load therefore may not be observed for several decades.

2. Reduced Cost

For a number of diseases, the costs of early detection and treatment have been determined and compared with the cost of later chronic illness. Cost reduction must be considered in two ways, however: first, by comparing the relative costs of doing and not doing a procedure (for example, mass TB screening *vs.* no screening) and, second, by comparing costs of different methods of screening assuming we have decided to screen at all. In the former case, we may elect to optimize the whole system and make decisions on tests on a cost basis; in the second, we may have made the basic decision on more emotional grounds and are then seeking the cheapest way out.

The one conclusion which is inescapable, however, is that cost reduction must be evaluated against total health care delivery costs. Does screening reduce the total bill? This again requires experiments on a sufficiently large scale and sophisticated statistical analysis.

3. Reduced Suffering and Unhappiness

Since the real aim in a health care–health research system is to reduce suffering and unhappiness, these parameters must enter into system judgments. The *patient reassurance* aspect of mass screening is an important part of its justification. Reassurance is in direct proportion to credibility. If the attending physician can say that the results of extensive valid testing are all negative and that no disease is present (and is correct in his conclusion), he has fulfilled one of his chief functions. Such reassurance (or a precise diagnosis) must rest on large amounts of statistical information. What the physician is doing is placing a patient in a category of patients whose history is known. This is possible only when the total number of patients studied is large and when precise multiparametric data are available from all.

The benefits to society, however, extend for a period of time much longer than the life span of an individual. Genetic screening therefore applies to present and subsequent generations. From the point of view of society, screening results may be divided roughly into those affecting this generation and those which may, through counseling and other means, affect the next. Since there is a great overlap, with many test results being the result of either genetic or environmental factors, it is not possible to make a clean separation of the two at the experimental level in all instances.

D. Benefits to Research, Medical, and Decision-Making Communities

1. Research Community Benefits

The allocation of training and research funds in the biomedical sciences presumably bears some relation to (1) the seriousness and extent of human disease, (2) the gaps in the application of existing knowledge, and (3) the explorability of areas of present ignorance. There is now little direct connection between research support decisions and real medical problems. This accounts for much failure to apply existing knowledge and the growing public realization that advances and implied advances described in the news media may often not reach the individual patient. One way to evaluate a research enterprise is to find out if it is solving the most important problems. This requires detailed information. If the objective is improvement of human health, then we must have a running status report. Do radiation and chemical pollutants produce mutations and disease in man? If a new drug produces delayed side-effects as serious as those produced by thalidomide, how soon would they be discovered? Basic decisions relating to the organization and conduct of research require a detailed running evaluation of the human condition at the biochemical level.

The credibility of the research community rests on the degree to which the gaps between the possible and the actual are closed. It has been an operating assumption in basic research that the application of new findings was easy and occurred automatically. This has obscured the fact that there are many problems unique to the mass application of any idea. The research facility required to develop a meaningful and useful screening system is an existing screening system which may be continually improved.

One of the major research advantages of a large-scale screen will be the identification of new diseases and disease groups. For example, the finding of "normal" individuals with a high uric acid level raises the question of whether this is a new genetic trait and also poses interesting problems to the biochemist: What enzyme differences are involved? Why are symptoms of gout absent? If a large number of human urine samples were screened for 50–100 metabolites, it is likely that dozens of new biochemical syndromes would be identified. These are, at least to the authors, challenging and interesting areas of investigation needed to help supply gaps in our information concerning the biochemistry of man.

2.   Medical Community Benefits

Biomedical research now makes possible a wide variety of sophisticated analyses. Which ones are useful? In the past, a new test has been popularized by advocates who usually either originated or modified the test. Sporadic evaluation follows, usually with equivocal results because of (1) poor patient choice (the specific advantageous use of the test is not recognized), (2) technical deficiency in the test (the scientific basis is good, but the test is not fully developed), (3) defective scientific basis. Some tests of dubious value such as the thymol turbidity test linger on for years. Also, a very long time is required for one test to supplant another, even though they detect the same disease. Any large-scale screening activity, carried on over a long period of time, will yield very useful information on the diagnostic value of specific tests and will improve the statistics on which medical decisions are based.

3.   Benefits to the Decision-Making Community

Increasingly, decisions relating to health care research and delivery will be group decisions based on group studies. This is true all the way from decisions on the dosage form of drugs to be manufactured to decisions on the form of patient records. They must increasingly be based on data, and the more specific the decision the more specific should be the data. Does a given pesticide produce an increased mutation rate in man? The answer is not obtained by committee introspection.

E. Ethical Considerations

The ethical problems include the following: Who has access to data? How much information should be available to the patient (or mother) and when? How should the patient be identified? What are the responsibilities of the screeners to see that some use is made of the data, especially when a treatable disease is uncovered? The last bears on the question of whether screening, genetic or otherwise, is an active or a passive activity. These are problems best solved by organizing a small screening program and determining experimentally how serious the problems are. Many of them have been faced already in TB, sickle cell anemia, and Tay-Sachs screening programs. The last two still fall in the special local community effort category and are unavailable to large segments of the population.

XV. CONCLUSIONS

Mutagenic factors which might affect man are generally present in the environment at low levels and in various combinations. While screening with experimental systems will be helpful in detecting obvious mutagens, in the final analysis only observations on man will satisfy concerns over the possible synergistic action of true mutagens and the role of peculiarities of human metabolism in enhancing the action of certain mutagens.

This review argues that, as of now, the preferred type of monitoring involves surveillance of a series of human proteins present in blood and metabolites in urine, for evidences of mutational change. No insurmountable obstacles exist to mass genetic monitoring, although a very large amount of research, development, and optimization remains to be done.

The overlap between techniques and data for mutation screening and screening for genetic diseases in the newborn is so large that it is inconceivable that these two activities should be organized or supported separately. It is therefore suggested that they be combined in a 5-year intermediate-scale screening program with the objective of embarking on massive studies at the end of that period.

XVI. REFERENCES

Abelev, G. I., Perova, S. D., Khramkova, N. I., Postnikova, Z. A., and Irlin, I. S. (1963), *Transplantation 1*, 174.
Abelev, G. I. (1971), *Advan. Cancer Res. 14*, 295.

Abrams, M. E. (1970), "Medical Computing," American Elsevier Publishing Co., New York.

Anderson, N. G. (1969a), *Anal. Biochem. 28*, 545.

Anderson, N. G. (1969b), *Science 166*, 317.

Anderson, N. G. (1970), *Am. J. Clin. Pathol. 53*, 778.

Anderson, N. G., and Caton, J. E. (1973), High resolution electrophoresis: Immunosubtraction, In preparation.

Anderson, N. G., and Willis, D. D. (1973), Centrifugal system for cell washing and agglutination detection, In preparation.

Ball, E. M. (1964), *in* "Plant Virology" (M. K. Corbett and H. D. Siler, eds.), University of Florida Press, Gainesville.

Bergmeyer, H. U. (1963), "Methods of Enzymatic Analysis," Academic Press, New York.

Bradley, T. B., Wohl, R. C., and Rieder, R. F. (1967), *Science 157*, 1581.

Brewer, G. J. (1970), *in* "An Introduction to Isozyme Techniques," pp. xii and 186, Academic Press, New York.

Burtis, C. A., Butts, W. C., and Rainey, W. T., Jr, (1970), *Amer. J. Clin. Pathol. 53*, 769–777.

Burtis, C. A., Johnson, W. F., Mailen, J. C., and Attrill, J. E. (1972), *Clin. Chem. 18*, 433.

Caton, J. E. (1972), Personal communication.

Caton, J. E., and Anderson, N. G. (1973), A centrifugal system for casting gradient gels, In press.

Caton, J. E., Willis, D. D., and Anderson, N. G. (1972), *Anal. Biochem. 46*, 232.

Cavalli-Sforza, L. L., and Bodmer, W. F. (1971), *in* "The Genetics of Human Populations," pp. xvi and 965, W. H. Freeman, San Francisco.

Chilcote, D. D., and Mrochek, J. E. (1971), *Clin. Chem. 17*, 751.

Chilcote, D. D., and Mrochek, J. E. (1972), *Clin. Chem. 18*, 778.

Chilcote, D. D., and Scott, C. D. (1971), *Chem. Instr. 3*, 113.

Chilson, O. P., Costello, L. A., and Kaplan, N. O. (1966), *Fed. Proc. 24*, (Suppl. 15), S-55.

Clegg, J. B., Weatherall, D. J., and Milner, P. F. (1971), *Nature 234*, 337.

Cleland, W. W. (1970), "Steady State Kinetics in the Enzymes" (P. D. Boyer, ed.), Vol. II, pp. 1–67. Academic Press, New York.

Cohen, M. M., and Bloom, A. D. (1971), *in* "Monitoring, Birth Defects and Environment" (E. B. Hook, D. T. Janerich, and I. H. Porter, eds.) pp. 249–272, Academic Press, New York.

Cohen, M. M., and Hirschhorn, K. (1971), *in* "Chemical Mutagens" (A. Hollaender, ed.), Vol. 2, pp. 515–534, Plenum Press, New York.

Coodley, E. L. (1970), "Diagnostic Enzymology," Lea and Febiger, Philadelphia.

Crow, J. F. (1971), *in* "Chemical Mutagens" (A. Hollaender, ed.), Vol. 2, pp. 591–605, Plenum Press, New York.

Davies, R. O., Morton, A. V., and Kalow, W. (1960), *Can. J. Biochem. Physiol. 38*, 545.

Dickerson, R. E. (1971), *J. Mol. Evolution 1*, 26.

Giblett, E. R. (1969), *in* "Genetic Markers in Human Blood," pp. xxvii and 629, Blackwell Scientific Publications, Oxford.

Gorodezki, V. K., Mikhailoff, V. I., and Rosenfeld, E. L. (1971), *Clin. Chim. Acta 34*, 67.

Guthrie, R., and Murphey, W. H. (1971), *in* "Phenylketonuria and Some Other Inborn Errors of Amino Acid Metabolism" (H. Bickel, F. P. Hudson, and L. I. Woolf, eds.), pp. 132–136, Georg Thieme Verlag, Stuttgart.

Harris, H. (1970), "The Principles of Human Biochemical Genetics," pp. ix and 328, North-Holland Publishing Co., Amsterdam.

Harris, H. (1971), *J. Med. Genet. 8,* 444.

Harris, H., and Whittaker, M. (1961), *Nature 191,* 496.

Hoffman, R., Brock, J., and Friemel, H. (1970), *Arch. Gynaekol. 208,* 266.

Hopkinson, D. A., and Harris, H. (1971), *Ann. Rev. Genet. 11,* 1.

Houston, I. B., and Veale, A. M. O. (1971), *Med. Lab. 9,* 30.

Hsia, D. Y. (1971), *in* "Early Diagnosis of Human Genetic Defects" (M. Harris, ed.), pp. 105–120, Government Printing Office, Washington, D.C.

Jellum, E., Stokke, O., and Eldjarn, L. (1971), *Scand. J. Clin. Lab. Invest. 27,* 273.

Jones, R. T., Brinhall, B., Huisman, T. H. J., Kleihauer, E., and Betke, K. (1966), *Science 154,* 1024.

Kalow, W., and Genest, K. (1957), *Can. J. Biochem. Physiol, 35,* 339.

Kimura, M. (1968*a*), *Genet. Res. 11,* 247.

Kimura, M. (1968*b*), *Nature 217,* 624.

Kimura, M. (1969), *Proc. Natl. Acad. Sci. 63,* 1181.

Kimura, M., and Ohta, T. (1971), *in* "Theoretical Aspects of Population Genetics," pp. ix and 219, Princeton University Press, Princeton, N.J.

Kinderlerer, J. L., Kilmartin, J. V., and Lehmann, H. (1971), *Lancet i,* 732.

King, J. L., and Jukes, T. H. (1969), *Science 164,* 788.

Levy, H. L., Shih, V. E., and MacCready, R. A. (1972), *in* "Early Diagnosis of Human Genetic Defects" (M. Harris, ed.), pp. 47–53, Government Printing Office, Washington, D.C.

Lüning, K. G., and Searle, A. G. (1971), *Mutation Res. 12,* 291.

Lyon, M. F., and Morris, T. (1966), *Genet. Res. (Cambridge) 7,* 12.

Mazur, P. (1970), *Science 168,* 939.

Medical Research Council (1956), *in* "The Hazards to Man of Nuclear and Allied Radiations," pp. vii and 128, Her Majesty's Stationery Office, London.

Medical Research Council (1960), "The Hazards to Man of Nuclear and Allied Radiations. A Second Report to the Medical Research Council," pp. vii and 154, Her Majesty's Stationery Office, London.

Melville, R. S., and Kinney, T. D. (1972), *Clin. Chem. 18,* 26.

Motulsky, A. G., and Yoshida, A., (1969), *in* "Biochemical Methods in Red Blood Cell Genetics" (J. J. Yunis, ed.), pp. 51–93, Academic Press, New York.

Mukai, T. (1970), *Drosophila Inform. Serv. 45,* 99.

National Academy of Sciences—National Research Council (1956), "The Biological Effects of Atomic Radiation. Summary Reports from a Study by the National Academy of Sciences," Washington, D.C.

Neel, J. V. (1957), *in* "Effect of Radiation on Human Heredity," pp. 139–150, World Health Organization, Geneva.

Neel, J. V. (1970), *Proc. Natl. Acad. Sci. 67,* 908.

Neel, J. V. (1971), *Perspectives Biol. Med. 13,* 522.

Neel, J. V. and Bloom, A. D. (1969), *Med. Clin. N. Am. 53, 1243.*

Posen, S. (1970), *Clin. Chem. 16, 71.*

Rosenfeld, E. L., Lukomskaya, L. S., and Gorodezky, V. K. (1965), *Clin. Chim. Acta 11,* 195.

Roth, M. (1971), *Meth. Biochem. Anal. 17,* 189.

Rowe, A. W. (1971), *Mech. Eng. 93(5),* 37.

Rowe, A. W., Eyster, E., and Kellner, A. (1968), *Cryobiology 5,* 119.

Russell, W. L. (1965), *Japan. J. Genet. 40,* 128.

Rutovitz, D., Farrow, A. S. J., Green, D. K., Hilditch, C. T., Paton, K. A., and Stein, R. (1970), *in* "Medical Computing" (M. E. Abrams, ed.), American Elsevier Publishing Co., New York.

Scheinberg, I. H. (1971), *in* "Monitoring, Birth Defects and Environment" (E. B. Hook, D. T. Janerich, and I. H. Porter, eds.), pp. 207–216, Academic Press, New York.

Schlager, G. (1972), *Mutation Res. 14,* 254.

Schlager, G., and Dickie, M. M. (1966), *Science 151,* 205.

Scott, C. D., and Mailen, J. C. (1972), *Clin. Chem. 18,* 749.

Scott, C.D., Jansen, J. M., and Pitt, W. W. (1969), *Am. J. Clin. Pathol. 53,* 739.

Scriver, C. (1972), *in* "Early Diagnosis of Human Genetic Defects" (M. Harris, ed.), pp. 87–98, Government Printing Office, Washington, D.C.

Shaw, M. W. (1972), *in* "Mutagenic Effects of Environmental Contaminants" (H. E. Sutton and M. I. Harris, eds.), Academic Press, New York.

Shih, V. E., Levy H. L., Karolkewicz, V., Houghton, S., Efron, M. L., Isselbacher, K. J., Beutler, E., and MacCready, R. A. (1971), *New Engl. J. Med. 284,* 753.

Smith, E. L. (1970), *in* "The Enzymes," 3rd ed. (P. D. Boyer, H. Lardy, and K. Myrbäck, eds.), Vol. 1, pp. 267–339, Academic Press, New York.

Stevenson, A C., and Kerr, C. B. (1967), *Mutation Res. 4,* 339.

Strickland, R. D. (1970), *Anal. Chem. 42,* 32R.

Sutton, H. E. (1971), *in* "Monitoring, Birth Defects and Environment" (E. B. Hook, D. T. Janerich, and I. H. Porter, eds.), pp. 237–248, Academic Press, New York.

Sutton, H. E. (1972), *in* "Mutagenic Effects of Environmental Contaminants" (H. E. Sutton and M. I. Harris, eds.), Academic Press, New York.

Thalhammer, O., Scheibenreiter, S., and Biedl, E. (1970), *Wien. Klin. Wschr. 82,* 1.

Tobari, Y. N., and Kojima, K. (1972), *Genetics 70,* 397.

United Nations (1962), "Report of the United Nations Scientific Committee on the Effects of Atomic Radiation," Seventeenth Session, Suppl. 16 (A/5216), United Nations Official Records, New York.

United Nations (1969), "Report of the United Nations Scientific Committee on the Effects of Atomic Radiation," Twenty-fourth Session, Suppl. 13 (A/7613), United Nations Official Records, New York.

Vasileiskii, S. S. (1965), *Byull. Eksperim. Med. 60,* 1266.

Veale, A. M. O., Lyon, I. C. T., and Houston, I. B. (1971), *New Zeal. Med. J. 74,* 83.

Vogel, F. (1970), *in* "Chemical Mutagenesis in Mammals and Man" (F. Vogel and G. Röhrborn, eds.), pp. 445–452, Springer-Verlag, Berlin.

Wallace, V. (1967), *Anal. Biochem. 20,* 411.

Weitkamp, L. (1971), *in* "Monitoring, Birth Defects and Environment" (E. B. Hook, D. T. Janerich, and I. H. Porter, eds.), pp. 217–232, Academic Press, New York.

Wilkinson, G. N. (1961), *Biochem. J. 80,* 324.

Williams, C. A., and Chase, M. W. (eds.) (1971), "Methods in Immunology and Immunochemistry," Vol. 3, p. 309, Academic Press, New York.

Williams, C. A., and Wemyss, C. T. (1961), *Ann. N.Y. Acad. Sci. 94,* 77.

Williams, G., Young, D. S., Stein, M. R., Cotlove, E., and Harris, E. K. (1970), *Clin. Chem. 16,* 1016.

Wright, G. L., Jr., Farrell, K. B., and Roberts, D. B. (1971), *Clin. Chim. Acta 32,* 285.

Young, D. S., Eploy, J. A., and Goldman, P. (1971), *Clin. Chem. 17,* 765.

Yunis, J. J. (1969), "Biochemical Methods in Red Blood Cell Genetics" (J. J. Yunis, ed.), Academic Press, New York.

Yunis, J. J. (1973), Genetic defects of erythrocyte enzymes, In press.

CHAPTER 29

Repair of Chemical Damage to Human DNA*

James D. Regan and R. B. Setlow

Carcinogenesis Program
Biology Division
Oak Ridge National Laboratory
Oak Ridge, Tennessee

I. INTRODUCTION

Our experiments and the substance of this chapter deal with chemical mutagens and chemical carcinogens and how they affect human cells *in vitro*. In particular, we assay damage to human DNA and the extent to which it is repaired after treatment with such chemicals. Thus, for our own purposes, we tend to consider all these chemicals as DNA-damaging agents, irrespective of their actions in other systems as mutagens, carcinogens, or both. To point up the importance of repair studies, we will first discuss the sequence of molecular events in hypothetical experiments that *do* assay mutation and/or carcinogenesis, and we will indicate at what point the nature of the primary lesion and the extent of its repair become matters of prime interest.

* Research jointly sponsored by the National Cancer Institute and the U.S. Atomic Energy Commission under contract with the Union Carbide Corporation.

II. SEQUENCE OF MOLECULAR EVENTS IN EXPERIMENTS WITH MUTAGENIC AND CARCINOGENIC AGENTS

A. Typical Test Systems

There have been many reports of experiments on the use of micro-organisms in studies of DNA-damaging agents. The reader is referred to other chapters in these volumes for review of these systems. In our experiments, we use human cells *in vitro,* and our discussion will be limited to mammalian-cell culture systems.

In a typical experiment on mammalian-cell mutagenesis (see Chu, Chapter 15 in Volume 2 of this series), mammalian cells with a particular phenotype, e.g., drug sensitivity, are exposed to a mutagenic agent and then assayed for colony-forming ability on medium containing the drug to which the cells are sensitive (e.g., 8-azaguanine; see Chu, Chapter 15, Volume 2, and Clive *et al.,* Chapter 27, this volume). What is assayed is the number of cells (i.e., colony formers) that have been mutated from drug sensitivity to drug resistance by the mutagen.

In a typical carcinogenesis experiment *in vitro,* hamster or mouse cells are treated with an agent, the appearance of transformed foci of growth is noted, and the cells arising from these foci are eventually tested for tumorigenicity by assay for tumor growth in a suitable host, i.e., mouse or hamster (Heidelberger, 1970; Di Paolo *et al.,* 1972).

B. Introduction of the Agent

Many DNA-damaging agents are insoluble in aqueous media, so it is often necessary to use some other solvent (dimethyl sulfoxide is used frequently) as a stock solution. Thus, if one is concerned about the operational dose of the agent, as opposed to the actual dose to the culture, stability of the agent in the stock solution presolvent must be considered. When the agent is placed into the culture medium, considerations of the stability of the agent in aqueous media and the effect of the solvent on the culture, depending on the quantities involved, may be of significance.

C. Effect of Serum in the Treatment Medium

With regard to alkylating and ethylating agents and certain other agents, the serum content of the medium plays a significant role in altering the actual dose to the cell, since these agents alkylate serum macromolecules, causing reduction in the effective dose of the agent to the cells.

D. Active and Nonactive Forms of Agents

A number of DNA-damaging agents are not available in their active forms or have both active and nonactive derivatives. Acetylaminofluorene (AAF) (Miller and Miller, 1969) is not an active (proximate) carcinogen, but its *N*-acetoxy derivative (*N*-acetoxy-AAF) is. The nitrosamines, well known as carcinogens *in vivo* (Lijinsky and Epstein, 1970), are inactive as mutagens or carcinogens unless preincubated with mouse liver microsomes (Malling, 1971) or reducing systems (Udenfriend *et al.*, 1954).

Thus one must consider whether a particular DNA-damaging agent is in the active form—*in vitro* treatment of transforming DNA (Mather *et al.*, 1971) may provide information on this question—or is converted to the active form by some cellular process. The efficiency and yield of such a process will thus be of significance.

E. Local Dose to the Cellular DNA

Thus, having considered all of the possible interactions of the agent under investigation with the various components of the test system (solvents, medium, serum, cell constituents, etc.), we come to the basic question of the local dose of the agent to the cellular DNA.

Roberts *et al.* (1971) have shown that in the case of alkylating agents the degree of cellular inactivation, in terms of colony-forming ability, is best correlated to the amount of alkylation, rather than the initial dose, of the agent. Another approach is to assay the damage to transforming DNA treated *in vitro*, as mentioned above (Mather *et al.*, 1971).

F. Induction and Repair of DNA Lesions

Because of the complex interactions described above, it is important to determine the number of lesions present in the test system at zero time. Although this is not operationally possible in all systems, one could measure the amount of alkylation at zero time (Roberts *et al.*, 1971), or in the case of damage induced by ultraviolet (UV) radiation one can ascertain the percentage of thymidine in cyclobutane pyrimidine dimers (Carrier and Setlow, 1971). With most chemical DNA-damaging agents, however, the nature of the primary lesion is obscure.

At this point, repair studies become important. Almost all DNA-damaging agents, whether electromagnetic (UV and ionizing radiation) or chemical, that have been investigated in the proper test system show evidence of a repair effect. We propose that it is possible to gain insight into the kind of lesion induced (in a general fashion) by a study of the nature of the molecular repair events that follow the initial DNA damage.

G. Residual (Unrepaired) Lesions

If we assume that DNA is the target molecule for carcinogenic action (by no means proved), as it is well accepted to be the target for mutagenic action, and that some amount of repair of the DNA lesions induced by carcinogens and mutagens does occur, then we are forced to conclude that the mutagenic action of mutagens and the carcinogenic action of carcinogens are attributable to residual or unrepaired lesions in the DNA. This suggests that mutagenic or carcinogenic events occur as a result of a molecular event that takes place before the repair is effected, either because the rate of repair is slow or because the repair system has been saturated. The final result of either situation is the appearance of mutant colonies in mammalian-cell mutation experiments and of foci of cells exhibiting malignant transformation in chemical carcinogenesis experiments.

III. WHAT CAN DNA REPAIR STUDIES TELL US?

A. Extent of Repair

In their simplest form, DNA repair experiments can be used to detect the extent of repair. The criterion may be differential split-dose response to colony formation, uptake of thymidine in G_0 cells after insult, or autoradiographic evidence of "non-S" synthesis. Under certain conditions (e.g., after UV irradiation), it is possible to assay directly for the induction of DNA lesions (i.e., pyrimidine dimers) at zero time and thereafter (Regan et al., 1968). In such cases, loss of the photoproduct from DNA is an indication not only of the existence of a repair system but also of its amplitude. Alkylation can be similarly quantitated and timed (Crathorn and Roberts, 1966).

B. Nature of the Repair Event

Ideally, we want to be able to interpret the amount of ^{3}H-thymidine taken up by G_0 lymphocytes, the number of silver grains in an autoradiogram, or the scintillation count of a labeled culture in terms of the nature of the repair events. Does heavy labeling mean a few lesions with extensive excision and insertion of new bases, or many lesions with little excision and insertion of new bases?

C. Nature of the Lesion

Our experiments suggest that the type of lesion induced can be described (in a general fashion) in terms of the repair sequence induced. For example, with some DNA-damaging agents DNA strand breaks are repaired rapidly with no extensive excision of bases. With others, there is a protracted period during which the DNA is nicked, and extensive excision of bases with concomitant extensive insertion of new bases takes place. The lesions induced by these agents must differ markedly. In the simplest terms, we might say that some agents either directly or indirectly cause a small break in the DNA, which is repaired rapidly with little base insertion, whereas others apparently produce a topological perturbation in the DNA, which is acted on by an endonuclease, causing a DNA strand break that is repaired by extensive excision and base insertion.

Before discussing our findings further, we must first review the various methods of studying repair, in particular with regard to the kinds of averages they measure.

IV. METHODS OF STUDYING REPAIR

A. Thymidine Uptake

The most convenient repair assay is thymidine uptake into G_0 lymphocytes (Evans and Norman, 1968). Since there is little semi-conservative DNA synthesis in such cells, and what synthesis there is may be further reduced by the addition of hydroxyurea, background incorporation is essentially zero and all the counts incorporated into an acid-insoluble cell fraction can be assumed to result from repair synthesis. This method is rapid and convenient, but by itself it provides no information as to the number or size of repaired regions; obviously, many repaired regions with a few base insertions or a few repaired regions with many base insertions would give similar results in terms of thymidine uptake. Nevertheless, for a quick measure of repair after insult by some suspected DNA-damaging agent, the method is ideal.

B. Unscheduled Synthesis

The concept of unscheduled synthesis and the autoradiographic experiment that measures it were introduced by Rasmussen and Painter (1964). Their publication presented the first evidence for DNA repair in mammalian cells. Cells are grown on slides or coverslips, irradiated or otherwise insulted, and then incubated with ^{3}H-thymidine (^{3}H-dThd) of high specific activity. Autoradiograms are prepared, and the resulting slides show two types of nuclei—heavily labeled nuclei, indicative of cells in the S phase, and, if some repair has occurred, lightly labeled nuclei containing five to 40 silver grains per nucleus above background. These lightly labeled nuclei are said to represent cells undergoing "unscheduled" synthesis. The method is sensitive, quantitative, and informative on an individual-cell basis. As with thymidine uptake, however, this autoradiographic method does not give a measure of the number or size of repaired regions in the DNA. There may also be some inaccuracy introduced (particularly when repair-deficient mutants are used) by a few cells just entering semiconservative synthesis. Usually, several days are required for development of the autoradiograms, although this delay can be avoided if ^{3}H-dThd of extremely high specific activity (greater than 40 Ci/mole) is used.

C. Repair Replication

It is important to have an experimental procedure that provides direct evidence that the new incorporation seen after insult is really in parental DNA (i.e., represents repair of previously existing DNA).

The "repair replication" measurement, a method involving differential centrifugation of density-labeled DNAs in cesium chloride, provides such a procedure. The method requires long centrifugations (more than 24 hr) and, often, rebanding of certain fractions from the primary centrifugation, but the procedure is extremely sensitive and informative (Rasmussen and Painter, 1966; Cleaver and Painter, 1968) in that it proves the new incorporation to be in parental DNA and indicates the number of bases inserted into the average repaired region (Painter and Young, 1972).

In repair replication experiments, cells are irradiated and then incubated in medium containing ^{3}H-5-bromodeoxyuridine (^{3}H-BrdUrd). After a suitable repair period the cells are harvested, and the DNA is extracted and centrifuged to equilibrium in cesium chloride. When the gradient fractions are collected, it is possible to follow the "new" DNA (semiconservatively synthesized during repair and at hybrid density) by ^{3}H counts. The "new" DNA is half substituted with bromouracil (BrUra) and sediments to a lower point on the gradient than the "old" or parental DNA, which is detected by absorbancy at 260 nm or by incorporation of a second radioactive label and is, of course, not density labeled. The appearance of ^{3}H counts in the "old" DNA region is indicative that repair has occurred and that the new incorporation is truly in the parental

DNA. Furthermore, if the initial number of DNA lesions or breaks induced by the insult (as in the case of ionizing radiation) and the specific activity of the ^{3}H-BrdUrd are known, it is possible to calculate the average number of base insertions per repaired region (Painter and Young 1972). A diagrammatic representation of the result of a typical repair replication experiment is shown in Fig. 1.

Billen's (1969) studies of DNA repair in *Escherichia coli* after X-irradiation suggest that most of the "repair synthesis" seen after irradiation is not base insertion into regions being repaired but represents new growing points in the DNA. On the basis of thymidine uptake experiments and the other repair assays discussed above, it is quite unlikely that this is the case for human cells. In G_0 peripheral leukocytes, for example, there is little semiconservative synthesis, and the hydroxyurea present would presumably inhibit any induction of new growing points.

D. Excision of UV-Induced Pyrimidine Dimers

In the case of UV irradiation, it is possible to make a direct measurement of the induction and subsequent disappearance of what is considered to be the UV-induced lesion of principal biological importance—namely, the pyrimidine dimer (Setlow and Carrier, 1964; Regan *et al.*, 1968; Carrier and Setlow, 1971). The radiochromatographic technique used has been reviewed in detail recently (Carrier and Setlow, 1971).

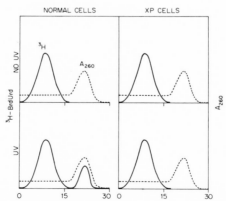

FIGURE 1. Assay of repair replication by cesium chloride centrifugation. Diagrammatic representation of the results of a typical experiment with normal and XP skin cells. Note "new" incorporation of ^{3}H-BrUra into region where "old" (parental) DNA sediments in normal cells (lower left panel); this is not present in XP cells (lower right panel).

E.   Photolysis of 5-Bromodeoxyuridine

Evidence presented by Lion (1970), Hutchison and Hales (1970), and Stephan *et al.* (1970) indicated that when *E. coli* DNA fully substituted with the thymidine analogue BrdUrd is exposed to 254-nm radiation, many chain breaks are produced, and additional bonds become sensitive to alkali. We have used their observations to devise a technique for assay of DNA repair. BrdUrd in the medium of irradiated cells is incorporated into repair-replicating regions as BrUra, and as a result 313-nm irradiation can be used to make the polynucleotide chains in the repaired regions sensitive to lysis by alkali. This wavelength (rather than 254 nm) is used because the molar absorbancy of thymine is about 100 times less than that of BrUra at 313-nm (Boyce and Setlow, 1963). Thus at 313-nm, breakage in BrUra-substituted DNA is much greater than in unsubstituted DNA. A diagrammatic outline of the method is shown in Fig. 2. We have employed this method to investigate differences in the repair capacities of normal and xeroderma pigmentosum (XP) cells (Regan *et al.*, 1971*a*), and we have shown how it can be used for intrauterine diagnosis of XP (Regan *et al.*, 1971*b*).

The BrUra photolysis assay is rapid and sensitive; it can detect one

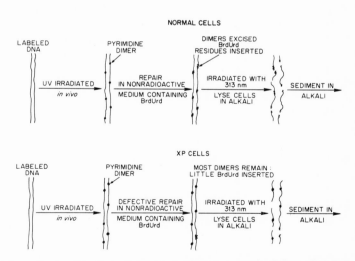

FIGURE 2.   Scheme for the detection of normal or defective repair by BrUra photolysis. The insult represented here is UV, resulting in pyrimidine dimers as the DNA lesion. In fact, however, any DNA-damaging agent can be investigated with this technique. For control purposes, both cell types are incubated in nonradioactive medium containing dThd rather than BrdUrd. (From Regan *et al.*, 1971*b*.)

repair event in 10^8 daltons of DNA. A particular advantage with this procedure is that it allows an estimate of two parameters of repair—the number of repaired regions, from the shift in molecular weight, and their size, from the sensitivity to the 313-nm radiation (the number of BrUra residues substituted is one-fourth the number of nucleotides). The change in the reciprocal of the weight-average molecular weight of the DNA, $\Delta(1/M_w)$, with increasing doses of 313-nm radiation is an indication of the size of the repaired regions; more BrUra residues provide a large target for the 313-nm radiation. $\Delta(1/M_w)$ equals $(1/M_w)_{\text{BrUra}}$ minus $(1/M_w)_{\text{dThd}}$, where $(1/M_w)_{\text{BrUra}}$ is the reciprocal of the weight-average molecular weight of the cells incubated after insult in BrdUrd and $(1/M_w)_{\text{dThd}}$ is the reciprocal of the weight-average molecular weight of the cells incubated in thymidine (dThd).

From a considerable number of experiments utilizing the BrUra photolysis assay on cells treated with UV and ionizing radiations, as well as a number of chemical agents that damage DNA, we have found that there are two rather distinct forms of repair in the DNA of human cells. Some agents cause a set of repair events typical of that seen after ionizing radiation (ionizing-type repair); others seem to induce repair typical of UV damage (UV-type repair). The rest of this chapter will be used to describe these two repair schemes, with UV and ionizing radiations as the model insults and BrUra photolysis as the assay. We will then describe our results with DNA-damaging chemicals that induce ionizing-type repair and compare them with those for chemicals that induce UV-type repair.

V. THE TWO FORMS OF REPAIR AS MEASURED BY BrUra PHOTOLYSIS

A. Ionizing-Type Repair

Human-cell DNA, as visualized on sucrose gradients by the technique of McGrath and Williams (1966), under the conditions of our experiments has a control weight-average molecular weight (M_w) of 1–2 $\times 10^8$ daltons. After 10 krad of ^{60}Co γ-rays (zero time, ice temperature), the M_w is 3×10^7; but after 60 min of incubation at 37°C, the DNA returns to its control (unirradiated) M_w (Fig. 3). If we perform this same experiment with BrdUrd in the medium during and after irradiation and then subject the cells to several doses of 313-nm radiation, we can derive an estimate of the number and size of the repaired regions. If we plot the data derived as the reciprocal of M_w $(1/M_w)$, the curve shown in Fig. 4 is obtained—a slowly rising linear slope. From this, we can calculate the size of the average repaired region. Our results indicate that about one BrUra residue or three to four nucleotides are inserted per radiation-induced chain break. This then is the picture of ionizing radiation repair: strand breaks with few nucleotide insertions rapidly repaired. Painter and Young (1972) reached a similar conclusion, using a repair replication technique.

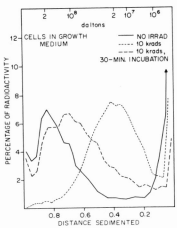

FIGURE 3.   DNA strand rejoining with time after 10 krad of ^{60}Co γ-rays, as observed on alkaline sucrose gradients.

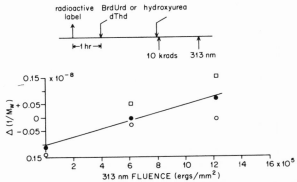

FIGURE 4.   Relationship of the reciprocal of the weight-average molecular weight $(1/M_w)$ to the dose of 313-nm radiation after DNA damage by 10 krad of ^{60}Co γ-rays and repair in BrdUrd.

B.   UV-Type Repair

Repair of human-cell DNA as observed in our assay after UV (254-nm) irradiation differs from ionizing-type repair in two essential aspects. First, the time involved for maximum incorporation of BrUra is 18–20 hr. Second, the size of the repaired regions (as determined by sensitivity to 313-nm radiation) appears to be equivalent to about 25 BrUra residues (100 nucleotides). Gradient profiles for a typical UV experiment with

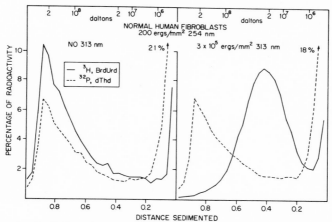

FIGURE 5.   Results of a BrUra photolysis assay for repair after 200 ergs/mm² of 254-nm UV. Note molecular weight shift induced by 3×10^5 ergs/mm² of 313-nm radiation, indicative of the extent of repair that has occurred in the BrdUrd-containing medium.

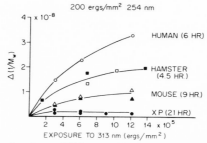

FIGURE 6.   Response to increasing doses of 313-nm radiation after repair in BrdUrd following 200 ergs/mm² of 254-nm UV as the initial insult. Response is indicated by change in molecular weight after 313-nm irradiation $[\Delta(1/M_w) = (1/M_w)_{\text{BrUra}} - (1/M_w)_{\text{dThd}}]$. Species vary considerably in the magnitude of repair taking place. Normal human cells repair extensively, hamster cells less, mouse cells even less, and XP cells only minimally.

normal human cells are shown in Fig. 5. Plotting these data as $1/M_w$, we see a curvilinear response that eventually, at higher doses of 313-nm radiation, reaches saturation (Fig. 6).

This curvilinear response to 313-nm radiation in these experiments, contrasted to that seen in γ-ray experiments, is an indication of target size for the 313-nm radiation. The number of BrUra residues per repaired region in the γ-ray experiments present a small target for 313-nm radiation. Saturation is impossible even at extremely high doses of 313-nm. In the case of UV-type repair, however, saturation is possible due to the extensive excision and BrUra substitution that has occurred in the average repaired region.

Thus, again comparing the two repair programs by the kinds of lesions induced and the enzymatic systems operating on the lesions (shown diagrammatically in Fig. 7), γ-irradiation induces small breaks that are rapidly repaired with a few nucleotides, generating in our assay a gradual, linear, unsaturated response to 313-nm radiation (Fig. 4), whereas UV irradiation induces larger breaks that are repaired by extensive excision of bases and extensive BrUra substitution, generating, in our assay, a curvilinear, eventually saturated response to 313-nm radiation (Fig. 6).

VI. CLASSIFICATION OF DNA-DAMAGING CHEMICAL AGENTS ACCORDING TO THE REPAIR SEQUENCE INDUCED

A. Ionizing-Type Repair after Chemical Damage

1. Ethyl Methanesulfonate (EMS)

Figure 8 shows a typical result in an analysis of DNA repair in human skin fibroblasts after treatment with EMS and subsequent incubation in BrdUrd. The $\Delta(1/M_w)$ curve generated by this procedure is qualitatively identical to a γ-ray curve (cf. Fig. 4). Quantitatively, there are more repaired lesions observed with this dose (10^{-2}M) of EMS than with 10 krad of γ-rays. Apparently, a number of small lesions are induced and quickly repaired with a few nucleotides and little BrUra insertion, presenting a target of minimal size for the 313-nm radiation. We would thus classify EMS as a typical ionizing-type DNA-damaging agent in terms of after-treatment repair events.

2. Methyl Methanesulfonate (MMS)

Our results with MMS are essentially the same as the results with EMS described above.

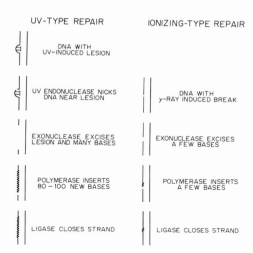

FIGURE 7. Diagrammatic representation of the two forms of repair in human cells after DNA damage, as visualized with the BrUra photolysis assay.

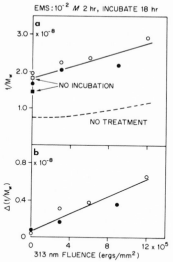

FIGURE 8. $\Delta(1/M_w)$ response to 313-nm radiation after EMS damage and repair in BrdUrd. Note similarity to γ-ray repair (Fig. 4).

3. Propane Sultone (PS)

PS is a highly reactive carcinogen that induces a high incidence of skin tumors in mice (Preussmann, 1968). In our assay, this agent induces repair events typical of ionizing radiation. Another highly reactive carcinogen, N-acetoxy-AAF, discussed below, produces markedly different repair events. Thus chemical carcinogenesis cannot be linked exclusively to a single sequence of after-treatment events.

B.   UV-Type Repair after Chemical Damage

1. N-Acetoxy-AAF

A report of our experiments with N-acetoxy-AAF has appeared elsewhere (Setlow and Regan, 1972). It is an agent that should hold considerable interest for photobiologists. We found that treatment of human skin-cell cultures with 0.7 µM N-acetoxy-AAF resulted (in the BrUra photolysis assay) in repair events *indistinguishable* from those induced by irradiation of the cells with 50 ergs/mm² of 254-nm UV (Fig. 9). In terms

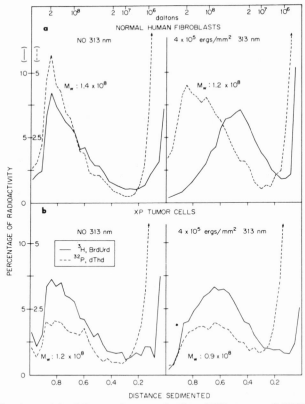

FIGURE 9.   Molecular-weight shifts seen in normal-cell and XP tumor-cell DNAs on alkaline sucrose gradients after treatment with the carcinogen N-acetoxy-AAF. Note similarity to UV repair (Fig. 5). (From Setlow and Regan, 1972.)

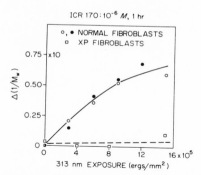

FIGURE 10. $\Delta(1/M_w)$ response to 313-nm radiation after treatment with ICR-170 in normal and XP cells. Note similarity to N-acetoxy-AAF (Fig. 11) and UV repair (Fig. 6).

of the number and size of the repaired regions, N-acetoxy-AAF clearly mimics UV. Further evidence for this mimicry is discussed below (Section VI).

2. ICR-170 [2-Methoxy-6-chloro-9-(3-[ethyl-2-chloroethyl]aminopropyla-mino)acridine Dihydrochloride]

ICR-170 is an acridine with one nitrogen mustard group. It has been employed as an antitumor agent (Creech et al., 1972). In our repair assay, this agent induced repair events similar although not identical to those induced by N-acetoxy-AAF (Fig. 10). ICR-170 differed from N-acetoxy-AAF in that the size of the repaired regions with ICR-170 was smaller (approximately ten BrUra residues) than those with N-acetoxy-AAF (approximately 25 BrUra residues). Nevertheless, ICR-170 clearly induces UV-type repair in normal cells and, like N-acetoxy-AAF, defective repair in XP cells (see Section VI).

VII. XERODERMA PIGMENTOSUM AND UV-TYPE REPAIR AFTER CHEMICAL DAMAGE TO DNA

Enough has been published on the hereditary disease xeroderma pigmentosum (XP), which causes extreme sensitivity to UV, that extensive explanation is not required here (Cleaver, 1968, 1969; Setlow et al., 1969; Regan et al., 1971a, b). When we observed the typical UV repair pattern in normal human cells after treatment with N-acetoxy-AAF, we began to examine XP cells for their repair response to this carcinogen. Again the results were typical of those seen in UV experiments with XP cells; i.e., there was little repair of the induced lesions (Fig. 11). We also examined the repair of UV-induced damage in a cell

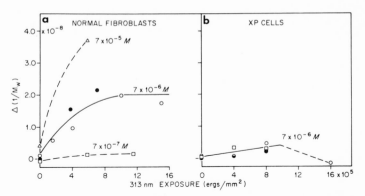

FIGURE 11. $\Delta(1/M_w)$ response to 313-nm after treatment with N-acetoxy-AAF in normal and XP cells. XP cells show only minimal repair after N-acetoxy-AAF treatment. (From Setlow and Regan, 1972.)

line developed from an amelanotic melanoma of an XP patient (previous studies had dealt only with nontumorous cells from XP patients) and found that the response of these cells, not surprisingly, was typical of other XP cells; i.e., they were deficient in repair after UV. Figure 9 shows the result of a repair assay in XP tumor cells after N-acetoxy-AAF treatment—again, highly deficient repair.

The results with N-acetoxy-AAF in XP cells led us to examine the response of XP cells to ICR-170, since UV-type repair was observed in normal cells after treatment with this agent. As expected, XP cells showed defective repair of lesions induced by ICR-170 (Fig. 10).

4-Nitroquinoline oxide (4-NQO) is a carcinogen that has been employed in repair studies in normal cells (Stich *et al.*, 1971; Stich and San, 1970) and in XP cells (Stich and San, 1971), which again show evidence of defective repair. Our results with 4-NQO in the BrUra photolysis assay are shown in Fig. 12. 4-NQO binds to DNA and can make it susceptible to photolysis as a result of photodynamic action (in much the same way as BrUra). Thus we can use 313-nm-induced strand breakage to assay the amount of 4-NQO bound to the DNA at various times after treatment. Figure 12a shows results from cells incubated only in dThd (no BrdUrd) after 4-NQO treatment. More DNA strand breaks are caused by 313-nm radiation at 90 min after treatment in XP cells than in the normal cells. This suggests that more 4-NQO remains bound to XP-cell DNA than to normal-cell DNA. After 18 hr, the number of breaks induced by 313-nm radiation in normal cells is nearly the same as that in untreated controls, suggesting that little or no 4-NQO is left in the DNA of these cells. The difference in the response of XP cells to 313-nm radiation at 90 min and at 18 hr may represent ionizing-type repair.

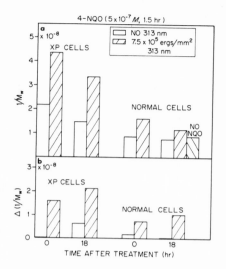

FIGURE 12. Magnitude of repair in normal and XP cells after treatment with 4-NQO.

This difference is probably not seen in normal cells without 313-nm radiation because the repair of such lesions in the normal cells may be essentially completed during the 90-min treatment period. Figure 12*b* shows data for strand breakage in cells incubated in BrdUrd, expressed as $\Delta(1/M_w)$. (The photodynamic effect of the 4-NQO is subtracted out by expressing the data in this fashion.) Both normal and XP cells show an increase in sensitivity to 313-nm radiation during the 18-hr repair period, suggesting some insertion of BrUra. The increase in sensitivity is somewhat less in XP than in normal cells. We interpret these results as indicating that 4-NQO induces at least two different types of lesions— perhaps an ionizing type and a UV type—and that XP cells are somewhat defective in repairing the UV-type lesions.

VIII. SUMMARY OF PRESENT INTERPRETATIONS

Table 1 presents a summary of our work in this field. Our present interpretation of the data we have collected can be stated succinctly:

1. Chemical agents that damage DNA can be classified into two major groups on the basis of the type of repair observed after treatment—those that induce ionizing-type repair and those that induce UV-type repair. One agent, 4-NQO, induces repair activity of both types.

TABLE 1. Summary of Damage and Repair Induced by Agents Used to Insult Human-Cell DNA

Agent	Dose or concentration	Length of treatment (min)	Breaks per 10^8 daltons after 10^6 ergs/mm^2 of 313-nm	BrUra residues per lesion	Type of repair
254-nm UV	200 ergs/mm^2		10	25	UV (normal and XP differ tenfold)
^{60}Co γ-rays	10 krad		0.6	1	Ionizing
N-Acetoxy-AAF	7 μM	60	4	40	UV (normal and XP differ tenfold)
4-NQO	0.5 μM	90	~2		UV and ionizing?
EMS	10 mM	120	~1.0		Ionizing
MMS	50 μM	5	~0.4		Ionizing?
Propane sultone	0.2 mM	120	~0.4		Ionizing?
ENU^a	0.1 mM	240	cross-linking		Essentially no repair
Mitomycin C	0.5 μg/ml	420	~0.4		Ionizing?
ICR-170	1 μM	60	1 cross-linking	~10	UV-like (normal and XP differ ~tenfold)
Urethane	0.1 mM	60			None—and no detectable damage

^{a}ENU, ethyl nitrosourea.

2. Chemicals that induce UV-type repair in *normal* cells induce lesions that are essentially irreparable in XP cells.

3. The term "UV-endonuclease" is something of a misnomer. The term is used to identify the endonuclease that makes a nick in a DNA strand next to a UV-induced pyrimidine dimer and thus initiates the repair process. Present evidence (Cleaver, 1969; Setlow *et al.*, 1969) suggests that this enzyme is defective in XP cells. From our data on defective repair of chemical damage to DNA in XP cells, it appears that it is not really a specific UV photoproduct, such as a dimer, that is recognized when repair is initiated (certainly neither *N*-acetoxy-AAF nor ICR-170 can induce pyrimidine dimers) but rather some topological perturbation caused by improper base stacking due to the intercalation of the chemical agent into the DNA. Such a lesion and the dimer-induced alteration in base stacking could be similar in topology and thus have similar affinities for the "UV-endonuclease."

X. ACKNOWLEDGMENTS

The authors are grateful to Drs. Edmund Klein and James L. German for providing access to patients and for helpful advice and discussion.

X. REFERENCES

Billen, D. (1969), *J. Bacteriol. 97*, 169.

Boyce, R., and Setlow, R. B. (1963), *Biochim. Biophys. Acta 68*, 446.

Carrier, W. L., and Setlow, R. B. (1971), *Meth. Enzymol. 21*, 230.

Cleaver, J. E. (1968), *Nature 218*, 652.

Cleaver, J. E. (1969), *Proc. Natl. Acad. Sci. 63*, 428.

Cleaver, J. E., and Painter, R. B. (1968), *Biochim. Biophys. Acta 161*, 552.

Crathorn, A. R., and Roberts, J. J. (1966), *Nature 211*, 150.

Creech, H. J., Preston, R. K., Peck, R. M., O'Connell, A. P., and Ames, B. N. (1972), *J. Med. Chem. 15*, 739.

Di Paolo, J. A., Nelson, K. A., and Donovan, P. J. (1972), *Nature 235*, 278.

Evans, R. A., and Norman, A. (1968), *Radiation Res. 36*, 287.

Heidelberger, C. (1970), *Europ. J. Cancer 6*, 161.

Hutchison, F. H., and Hales, H. G. (1970), *J. Mol. Biol. 50*, 59.

Lijinsky, W., and Epstein, S. S. (1970), *Nature 225*, 21.

Lion, M. B. (1970), *Biochim. Biophys. Acta 209*, 24.

Malling, H. V. (1971), *Mutation Res. 13*, 425.

Mather, V. M., Lesko, S. A., Straat, P. A., and Ts'o, P. O. P. (1971), *J. Bacteriol. 108*, 202.

McGrath, R. A., and Williams, R. W. (1966), *Nature 212*, 534.

Miller, J. A., and Miller, E. C. (1969), *Progr. Exptl. Tumor Res. 11*, 273.

Painter, R. B., and Young, B. R. (1972), *Mutation Res. 14*, 225.

Preussmann, R. (1968), *Food Cosmet. Toxicol., 6*, 576.

Rasmussen, R. E., and Painter, R. B. (1964), *Nature 203*, 1360.

Rasmussen, R. E., and Painter, R. B. (1966), *J. Cell Biol. 29*, 11.

Regan, J. D., Trosko, J. E., and Carrier, W. L. (1968), *Biophys. J. 8*, 319.

Regan, J. D., Setlow, R. B., and Ley, R. D. (1971a), *Proc. Natl. Acad. Sci. 68*, 708.

Regan, J. D., Setlow, R. B., Kaback, M. M., Howell, R. R., Klein, E., and Burgess, G. (1971b), *Science 174*, 150.

Roberts, J. J., Pascoe, J. M., Plant, J. E., Sturrock, J. E., and Crathorn, A. R. (1971) *Chem.-Biol. Interactions 3*, 29

Setlow, R. B., and Carrier, W. L. (1964), *Proc. Natl. Acad. Sci. 51*, 226.

Setlow, R. B., and Regan, J. D. (1972), *Biochem. Biophys. Res. Commun. 46*, 1019.

Setlow, R. B., Regan, J. D., German, J., and Carrier, W. L. (1969), *Proc. Natl. Acad. Sci. 64*, 1035.

Stephan, G., Mitonberger, H. G., and Hotz, G. (1970), *Z. Naturforsch. 25B*, 1037.

Stich, H. F., and San, R. H. C. (1970), *Mutation Res. 10*, 389.

Stich, H. F., and San, R. H. C. (1971), *Mutation Res. 13*, 279.

Stich, H. F., San, R. H. C., and Kawazoe, Y. (1971), *Nature 229*, 416.

Udenfriend, S., Clark, C. T., Axelrod, J., and Brodie, B. B. (1954), *J. Biol. Chem. 208*, 731.

CHAPTER 30

Tradescantia Stamen Hairs: A Radiobiological Test System Applicable to Chemical Mutagenesis*

A. G. Underbrink,† L. A. Schairer, and A. H. Sparrow

Biology Department
Brookhaven National Laboratory
Upton, New York

I. INTRODUCTION

Several species in the family Commelinaceae, of which *Tradescantia* is a member, have features particularly well suited for certain radiobiological studies. Effects produced by ionizing radiations which are easily studied include (1) chromosome aberrations in microspores, root tips, ovaries, and stamen hairs; (2) somatic mutations in petals and stamen hairs in clones heterozygous for flower color; (3) pollen abortion; (4) loss of reproductive integrity in stamen hairs; and (5) whole plant or seedling death. The stamen hairs of *Tradescantia* clone 02 have also proved

* This research was carried out at Brookhaven National Laboratory under the auspices of the U.S. Atomic Energy Commission. It was supported in part by the Bureau of Radiological Health, U.S. Public Health Service Research Grant EC-00074, PHS Research Grant CA-12536 from the National Cancer Institute, and the National Aeronautics and Space Administration (Purchase Order A-44246A).
† Present affiliation: Department of Radiology, Radiological Research Laboratories, College of Physicians and Surgeons, Columbia University, New York, New York. Mailing address: Biology Department, Brookhaven National Laboratory, Upton, New York 11973.

to be sensitive to radiation-induced mutations at doses in the millirad region. [1]

The techniques for using the stamen-hair system have been developed over the past several years specifically for radiobiological studies. However, preliminary experiments indicate that these techniques are applicable to chemical mutagen studies in either a gaseous or liquid state. Several years ago, Smith and Lotfy [2] showed that certain gaseous chemicals, specifically ethylene oxide, ketene, and methyl chloride, produced chromosome aberrations in pollen tubes of *Tradescantia paludosa* cultured to metaphase. More recent experiments have also employed microspores. [3]

Many different plant systems can be used for chemical mutagenicity, as recently reviewed. [4–10] However, the efficiency of different plants can vary greatly. [11] One advantage of using *Tradescantia* is that it can be grown in the test environment for long periods of time and thus can act as a mutagen monitor. Periodic checking of the heterozygous flower petals or stamen hairs for somatic mutations can then be used to indicate whether the environment is mutagenic and, if so, to what extent. Initially, the effects induced can be compared with those resulting from exposure to X- or γ-rays both as to type and as to level of effect. However, as with radiation-induced aberrations, the lower the mutagen dose, the more flowers must be examined to show significant increases in somatic aberrations over control levels. [11]

The potential of the *Tradescantia* stamen-hair system for use as a monitor for chemical mutagenicity studies was indicated by a recent incident at Brookhaven. [12] Stock plants of *Tradescantia* clone 02, growing in a controlled environment room, were accidentally exposed to a gaseous chemical mutagen. The details of this incident have been described elsewhere. [12] In brief, vapors from various chemicals in use at the biology department were vented through a hood to a flue on top of the building. Investigation suggested that prevailing atmospheric conditions at the assumed time of exposure created a downdraft, during which time the chemical agents entered an air-conditioning intake duct leading eventually into the growth room. Although the mutagen could not be definitely identified, one of the chemicals employed at this time was diazomethane, which is known to be mutagenic. [13] This suspicion was somewhat strengthened when it was shown that immersion of *Tradescantia* clone 02 inflorescences in a 1.0 mM aqueous solution of nitrosoguanidine (which has degradation fractions including diazomethane [14]) produced results similar to those observed from the unknown mutagen in the growth chamber. [12]

Stamen hairs from the plants exposed to the unknown mutagenic agent contained an abundance of all the types of somatic mutations and aberrations which occur in *Tradeecantia* following irradiation. Comparison of the aberration rate of the plants in the growth chamber with that of X-irradiated material indicated that the mutagen produced responses approximately equivalent to a dose of 90 rads of X-rays. [12]

A subsequent pilot experiment revealed that *Tradescantia* inflorescences were sensitive to a commercial insecticide whose active ingredient consisted largely of 2, 2-dichlorovinyl dimethyl phosphate. This agent, however, produced mostly bud blasting with little or no mutational response. [15]

Only preliminary experiments have been undertaken to date, to evaluate the potential usefulness of the stamen-hair system for chemical-mutagen studies, but development of the system was continued to further improve its radiobiological usefulness. The established techniques used in physical mutagen testing will be described here.

Since *T. paludosa* was first used by Sax [16] to study induced chromosome aberrations, many radiobiological experiments have utilized various species or varieties of *Tradescantia* [1,3,16–62] or other commelinaceous plants. [63] However, no other single taxon so far studied offers such versatility as *Tradescantia* clone 02. This versatility was exemplified by the number of plant parts used in the Biosatellite II experiment. [32–34,44,47,48] Following the 2-day orbital flight, plants which were irradiated while in free flight (or their unirradiated flight controls) were analyzed for aberrations in petals, stamen hairs, microspores, pollen, megaspores, and root tips. Presumably, this versatility will prove advantageous in the realm of chemical mutagenesis as well.

Although many tissues or organs of this plant may be used efficiently in various experimental procedures (Table 1), this chapter will be concerned primarily with a detailed description of the procedures currently being used in handling the *Tradescantia* clone 02 stamen-hair system. Certain other *Tradescantia* clones, heterozygous for flower color, may be handled in the same way. [15,26,29,31]

Tradescantia clone 02 is a diploid plant ($2n = 12$), heterozygous for flower color, thought to be the result of a natural cross between blue- and pink-flowered forms or species. It is a perennial, herbaceous plant with narrow tapering leaves not unlike some of the medium-size grasses, and it reaches a height of about 2 ft (60 cm) (Fig. 1). Clone 02 has not been described botanically in the literature, but descriptions of the morphology and floral anatomy of *T. paludosa*, which is similar to clone 02, have been given by Gunckel *et al.*, [24] Parchman, [64] and Savage. [43]

Clone 02 was originally discovered in 1958 by one of the authors (A.H.S.) growing outdoors in a collection of *Tradescantia* assembled by Professor W. V. Brown at the University of Texas at Austin. Its position between regular rows of plants suggested that it originated as a seedling. The parental species are not known, but *T. occidentalis* and *T. ohiensis*

TABLE 1. Various Plant Parts, End Points, and Characters Usable as Test Systems in *Tradescantia* Clone 02

Plant part	End point	Character of feature scored
Whole plant (or seedlings)	Growth inhibition	Weight, height, etc.
	Lethality	Dead plants
Mature flower	Flower production	Flower number
Developing flower buds	Bud blasting	Number of dead buds
Petals	Somatic aberrations	Pink cells or sectors
	Growth inhibition	Petal size or area
	Chromosome breakage	Chromosome aberrations
Stamens	Reduced hair number	Number of stamen hairs/ anther
Mature stamen hairs	Loss of reproductive integrity	Number of cells/hair
	Chromosome breakage	Micronuclei
Mature stamen-hair cells	Somatic aberrations	Pink normal-size cells
		Pink giant cells
		Pink dwarf cells
		Colorless normal-size cells
		Colorless giant cells
		Colorless dwarf cells
		Blue giant cells
		Blue dwarf cells
Developing stamen hairs	Chromosome breakage	Chromosome aberrations
		Micronuclei
Meiocytes	Chromosome breakage	Chromosome aberrations
		Micronuclei
Mature pollen	Cell survival	Normal and aborted pollen
		Pollen germination
	Chromosome breakage	Micronuclei
	Polyploidy	Nuclear volume
Germinating pollen	Chromosome breakage	Chromosome aberrations
Ovary	Embryo sac abortion	Nondevelopment
	Spindle aberrations	Nuclear misorientation
	Chromosome breakage	Chromosome aberrations
		Micronuclei
Root tips	Growth inhibition	Weight, length, reduced cell number
	Chromosome breakage	Chromosome aberrations
		Micronuclei

have been suggested.[36] *Tradescantia* clone 02 has appeared in the literature as *T. occidentalis*, *T. ohiensis*, or an interspecific hybrid of *T. occidentalis* × *T. ohiensis*.[36,65]

Techniques for the successful and efficient radiobiological use of the stamen-hair system have been evolved during the last decade, but its

FIGURE 1. Mature stock plant of *Tradescantia* clone 02, in a 6-inch pot, showing its characteristic growth habit.

first extensive use was as a test system for mutation induction in the Biosatellite program. [32–34,44,47,48] It has been used in RBE experiments with monoenergetic neutrons, [1,50,51,53–60,66] and recently the authors have used high-energy nitrogen ion beams. Those workers credited with helping to develop various aspects of the techniques now in use are listed in the Acknowledgments. Unless these techniques are employed in a similar fashion by other experimenters, it will be impossible to intercompare results. It was decided, therefore, to compile a detailed description of techniques used routinely at Brookhaven and to a limited extent elsewhere to assist others in using the stamen-hair system efficiently.

Hairs of various types are not rare in plants, [67,68] but they generally consist of single or few cells, and most have not proved useful for radiobiological experiments. The comparatively uncommon stamen hairs, on the other hand, consist of long filaments of contiguous single cells (Fig. 2B) in which ionizing radiation or chemicals can induce several types and combinations of mutations or aberrations. Some aberrations,

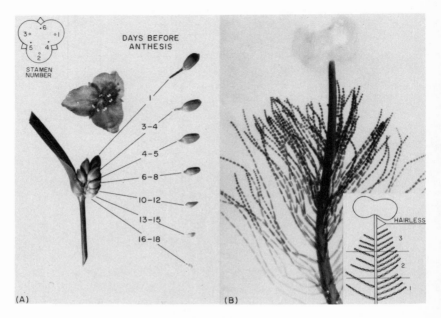

FIGURE 2. A, Inflorescence from clone 02 illustrating progressive increases in bud size with development from the most basal to the fully developed distal buds and an open flower. Mature buds (5–7 days before anthesis) may be removed prior to experimentation if required. Inset: Numbering system used to distinguish between antipetalous (odd numbers) and antisepalous (even numbers) stamens. B, Typical stamen with associated stamen hairs as they appear when properly prepared for scoring. Inset: Diagram of a typical stamen divided into three equal hair-bearing regions.

such as the change from the wild-type blue color to pink in clone 02, are considered to be true somatic mutations in the classical sense, as the plant is heterozygous for blue and pink flower color. [36,41]

Features that make the stamen-hair system highly advantageous for radiobiological work include the following: (1) Stamen hairs appear to be produced largely by repeated division of the terminal or subterminal cells with only a few interstitial divisions. Therefore, at certain developmental stages, they appear to have essentially a one- or two-celled meristem. [15,21,27,29,30,41] This feature favors the study of loss of reproductive integrity (hair stunting) after irradiation. Thus, survival curves and mutation rates can be compared reasonably well with data from single-celled systems or cultures of other organisms. [30,41,50,69] (2) Individual inflorescences contain a compact cluster of small flower buds, which permits uniform irradiation even with small beams. [53,59,70] Therefore, stamen hairs can be used for very accurate RBE determinations using neutrons of various energies or other radiations, [21–23,26,54,58–60] because the cuttings are small and hardy enough to permit the use of an adequate population in facilities with very limited space. (3) Large cell populations with well-defined cell lineages can be analyzed with regard to the location of specific kinds of aberrations. [27,39] (4) The handling of the material and the scoring of physical or chemical effects are readily learned and easily performed. [27,29,47,48,59] (5) A relatively short experimental period is required, although this time can vary. If rooted cuttings are chosen for the experiment, then 10 days are required for rooting before treatment. Following treatment, effects will not occur for a post-treatment period of about 1 week. Following this elapsed time, the actual scoring period may vary from 5 days to 3 weeks or longer, depending on the nature of information desired [1,27,41,47,48,54,55,59,60] and temperature used. [49] (6) The growth habits and other natural variables inherent in this plant are well known from previous work. [27,29,30,32,34–41,47–49,57–59,64] The plants grow well at relatively low light levels in growth chambers, providing a continuous year-around supply of flower buds at all stages of development. (7) *Tradescantia* clone 02 has a high radiosensitivity [29,30,32, 34–41,47–50,54–60]; its somatic cells have the same range of sensitivity as mammalian cells for loss of reproductive integrity but are probably more sensitive for somatic mutations. A significant yield of mutations has resulted from single acute doses of 10 mrad of 0.43-MeV neutrons and 0.25 rad of X-rays [1] and after chronic γ-ray exposures as low as 33 mR/hr for about 30 days. [49] (8) There are many species of the genus *Tradescantia* which differ in nuclear and chromosome volume, chromosome number, DNA content per cell, and degree of ploidy so that the influence of these variables may be investigated also. [17,26–28,64] (9) Genetic variability is minimal when the plants are propagated asexually by cuttings, but seed progeny of clone 02 are highly variable.

II. TECHNIQUES AND PROCEDURES

A. Cultivation of Plants

Tradescantia clone 02 is easily cultivated in an ordinary greenhouse. However, the plants are sensitive to certain environmental fluctuations, as demonstrated by increased bud blasting and aberration frequency with varying or unsuitable environmental conditions such as temperature, [49] humidity, heat shock, presence of various gases. [32,44] For this reason, the plants we use for most experiments are grown in controlled-environmental growth chambers or are grown with standard care under greenhouse conditions and then acclimatized in a growth chamber at least 2 weeks prior to taking cuttings. The chambers routinely used at Brookhaven for clone 02 are both walk-in and reach-in controlled-environmental growth chambers (Environmental Growth Chambers, Chagrin Falls, Ohio), but other types have also been used. Our plants are grown routinely under an 18-hr day, 1800 ± 400 ft-candle light regime using Sylvania fluorescent lights FR96 T12/CW/VHO/135° supplemented by about 10% incandescent light. Humidity is maintained between 60 and 65%. The temperature is maintained at $20 \pm 1°C$ for day and $18 \pm 1°C$ during the night cycle. However, other temperature and light regimes may also be used.[39] The air intakes to these chambers have been equipped with activated charcoal filters because of the known or suspected sensitivity of *Tradescantia* to various gases or pollutants that might enter.[12]

Stock plants are usually grown in 6-inch clay pots using a potting mix of sand, soil, and peat moss (1:1:1) with no added trace elements. Small numbers of stock plants, however, have been grown successfully in liquid culture[52,71] for use in mineral deficiency studies.

Routine maintenance of the stock plants includes conventional insect spray programs, but sprays must be carefully selected since some pesticide sprays are known to be mutagenic.[12,72] Weekly applications of Peters' liquid fertilizer (20-20-20 NPK, from Robert B. Peters Co., Inc., Allentown, Pa.) and removal of old flowers and/or old inflorescences are recommended. The removal of old inflorescences above a node forces new shoots to develop at the node. This assures a continuous supply of new inflorescences about 4–6 weeks later and reduces the possibility of seedling contamination of the stocks in this self-fertile clone. The possibility of selfing or crossing can be reduced further by excluding bees and other insects from the growing areas.

Every 6–8 months, the stock plants should be divided, the main stems cut back leaving two or three nodes above soil level, any seedlings removed, and the plants repotted in clean 6-inch clay pots. The stock plants become well established and are normally ready for a harvest of inflorescences about 6–8 weeks after transplanting or division.

B. Screening the Plants

This hybrid clone of *Tradescantia* has the usual range of biological variability from plant to plant for the various end points scored during an experiment. To minimize this variability and to maintain stock plants having a uniformly low spontaneous aberration frequency, it is desirable to screen the entire group of stock plants for aberrations every $1\frac{1}{2}$–2 years. An alternative is to reclone from a single selected plant.

The screening is accomplished by selecting one flower from each of five or six stalks on each stock plant. Two freshly picked petals from each flower are placed dry between two 40 by 80 mm glass slides and scored for the number of pink flecks or sectors per petal under a dissecting stereoscope (magnification approximately 25×). Small or even single-celled sectors can be scored. Pink mutations, in our material, occur in petals with a mean spontaneous frequency of about three per petal and therefore represent a sensitive system for screening plants. This petal system is limited for mutation studies, however, by the comparative difficulty in determining the number of cells per mutant sector or per petal and hence quantifying aberrant events on a per cell basis. (Mericle and Mericle [35–39] and Sparrow *et al.* [47] may be consulted for the techniques and uses of petals.)

If the number of pink sectors exceeds the mean spontaneous rate by a factor of 3 in the screening procedure, the plant is discarded or further scoring is undertaken to determine whether the frequency stays above the normal range.

Whenever possible, seedlings should be discarded. Seedlings arising from either self- or cross-pollinated seeds may not be heterozygous for flower color and hence will not show the spectrum of mutations characteristic of clone 02. Seedlings which are heterozygous may respond differently to mutagen treatment. [15] It is thus important to avoid contamination of the standard clone 02 by new seedlings. Occasionally, older plants may tend to lose their normal growth characteristics and assume a spindly habit and become overgrown with many nonflowering basal shoots. The reasons for this sporadic phenomenon are not clear, but these plants likewise are discarded.

C. Selection and Maintenance of Cuttings for Experimental Purposes

Vigorous shoots (4–8 inches long) consisting of two to four nodes bearing young inflorescences at a uniform stage of development should be selected for cuttings. It is desirable to select inflorescences whose most mature buds are within 2 days of anthesis because (1) the cuttings are more vigorous, (2) subsequent flower production is better, (3) some reduction in the number of hairs per stamen and of cells per hair may occur in older inflorescences during the last weeks of inflorescence activity, [38] and (4) the radiosensitivity of the inflorescence with respect to somatic mutation may also tend to decrease as the inflorescence becomes a few weeks old. [39] It is not determined as yet whether sensitivity to chemical mutagens also declines.

The cuttings may be either axillary shoots or from primary shoots cut off at any appropriate node. Axillary cuttings root faster than nodal cuttings (in less than 7 *vs.* 7–10 days) because root primordia are already formed at the basal node on the axillary cuttings. However, both types root quite well, with a rooting average of nearly 100% for axillary cuttings and about 80% for nodal cuttings. Typical cuttings are shown in Fig. 3.

Somatic mutation rates are essentially the same whether flowers are from rooted or unrooted cuttings. However, flower production declines more rapidly or occasionally may even stop when the cuttings are not rooted (Fig. 4). Peak days for mutation frequencies also may change. For rooting, any excess stem and leaf material is removed from the area of the basal node and the base of each cutting is dipped in water and dusted with a commercial root-inducing substance such as Hormodin (0.1% indolebutyric acid). The use of a fungicide does not seem necessary. The cuttings are placed in a medium such as vermiculite (horticultural grade) or Perlite (Coralux Perlite Corp. of New Jersey) saturated with Hoagland's No.2 solution [73] in shallow plastic containers with holes in the bottom for drainage. Rooting may be hastened if the cuttings are placed in a high-humidity propagation chamber and given bottom heat (27–28°C). The temperature in the root zone should be monitored, as excessive heat damages the roots. Adequate rooting should take place in about 10 days, and treatments may be given on the eleventh day (or later if desired).

FIGURE 3. Cuttings, both rooted and unrooted, of a size convenient for experimentation. Small nutrient tubes with caps may be used if experimentation lasts longer than a day.

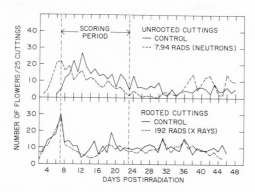

FIGURE 4. Flower yields expected from 25 cuttings during the period of experimentation, for both rooted and unrooted cuttings. The flower yield is generally higher for longer periods in rooted cuttings, but at these radiation doses there is no clear relation between dose and flower production.

After irradiation or other treatments, lots of up to 25 cuttings may be placed in 400 ml beakers containing aerated Hoagland's solution. Each beaker is aerated using an aquarium pump attached to drawn glass tube aerators. The Hoagland's solution should be changed once a week, and algal growth can be retarded by wrapping each beaker in black paper.

D. The Flower

Using normal daylight from a north window, the blue-purple flower color corresponds well to the full hue or lighter tints of Color No. 41, Lobelia Blue, and the purplish-red (or pink) colors of the mutants fall most often within the range of tints to full hue to Color No. 32, Petunia Purple,[30,36] as described in the "Horticulture Colour Chart."[74]

These flower colors and the color of the stamen hairs may vary greatly in intensity. These intensity changes are probably due to variations in the growing conditions; i.e., it was observed that petal color appears intensified in plants grown at cooler temperatures and more dilute at higher temperatures. Occasionally, faintly colored flowers or stamen hairs show up among deeply hued flowers in both experimental and control cuttings. In most cases, these should be discarded.

Bud development in the inflorescence of *Tradescantia* clone 02 proceeds in an orderly and continuous sequence from bud initiation to anthesis, with the youngest buds at the base of the inflorescence[38,43] (Figs. 2A and 3). The development is very similar to that described in detail for *T. paludosa*.[24] An inflorescence on a potted plant may produce flowers and initiate new buds for over 5 months, ultimately producing as many as 140 or more flowers if the flowers are picked daily.[38]

In rooted cuttings, grown as described above, flowers are produced on an average of about one every other day per inflorescence over a 23-day scoring period. Thus, in planning experiments with clone 02, enough cuttings should be used to ensure an adequate supply of flowers, bearing in mind that flower production usually decreases during the post-treatment scoring period. This decrease is usually more severe in fresh cuttings than in rooted cuttings (Fig. 4). Except at rather high radiation doses (for clone 02 about 400 rads of X-rays or 25 rads of 0.43-MeV neutrons) and rather late in the scoring period, the decrease in flower production appears to be relatively dose independent.

A flower lasts only a few hours if left on the cutting at room temperature. For this reason, flowers should be picked shortly after the buds open in the morning and stored on moist filter paper in appropriately labeled covered containers (petri dishes, plastic egg containers, etc.) at 4–6°C. Thus stored, the stamen hairs should remain scorable for about 2 days if picked when the buds first open. The flowers, however, should be fully open for convenience of removing individual stamens, because closed buds will not open while being refrigerated nor will the filament elongate to its full length.[66] Since the flowers open fully about 4 hr after the daily light cycle has started, the time of day at which the flowers open may be changed for convenience by advancing or retarding the beginning of the 18-hr light cycle. After high radiation exposures, some flowers may fail to open properly. However, the stamens usually can be dissected out and scored.

The flower has three blue petals alternately arranged with three sepals. There are six stamens (three antipetalous and three antisepalous). The three antipetalous stamens are somewhat longer and may contain more than 8% more hairs than the antisepalous stamens. [41,66] Originally, it was thought that the antipetalous stamens might respond differently to irradiation than the antisepalous stamens. To test this relationship, each stamen was numbered so that any information pertaining to particular stamens could be extracted during computer analysis. In numbering the stamens, an arbitrarily chosen antipetalous stamen is given the number 1, and then in a clockwise fashion the next antipetalous stamen is 2, the next 3. The first antisepalous stamen, between Nos. 1 and 2 antipetalous stamens, is number 4, etc. (Fig. 2A). Present evidence indicates that the aberration frequency in our material is identical in both types of stamens when analyzed on a per hair basis.

E. Flower Production Records

Daily records of each cutting may be kept to determine whether particular cuttings have abnormally high aberration rates compared to other cuttings which bloomed on that or later days. Similarly, by knowing which cutting is used, i.e., by tagging individual cuttings (Fig. 5), one can get an estimate of the age of the inflorescence. For example, flowers produced later on an inflorescence may have different mutation rates and/or more stunted or fewer hairs than those produced on earlier flowers.[38]

Prior to irradiation, all flowers and mature buds (those within 5–7 days of opening under our growth conditions and having completed all cell division) are removed. Older buds are too mature to yield a genetic

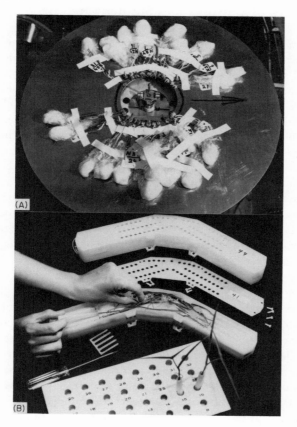

FIGURE 5. A, Cuttings of *Tradescantia* in position for neutron irradiation. The water-cooled tritium-impregnated target is in the center. The inflorescences are grouped around it in such a position that the buds are uniformly irradiated with respect to dose and neutron energy. The base of each cutting is enclosed with moist cotton and Saran Wrap (thin sheet plastic). Individual cuttings may be tagged if flower production records are kept. (From Underbrink *et al.*[57]) B, Close-up of Biosatellite II hardware being loaded with *Tradescantia* cuttings in preparation for 2-day orbital flight. (From Sparrow *et al.*[47])

response to experimental treatment in the stamen hairs, and their careful removal does not appear to cause adverse effects. The removal of mature buds results in an inflorescence small enough to be grouped with several others for experimentation, which is helpful when the target space is very limited (Fig. 5A). If the experimenter so wishes, the buds that are removed may be fixed in Newcomer's solution or some other appropriate fixative, transferred to 70% ethanol, and stored for future reference to estimate the stage of development of the pollen or treated buds.

F. Numbers of Cuttings and Stamens Required

In experimental situations, the number of stamens to be scored is limited by the technical personnel and facilities available or by the number of stock plants. However, it is particularly important to collect all possible data from such expensive and nonrepeatable experiments as Biosatellite II.[17,32–34,47,48] In that experiment, all of the stamen hairs from all of the flowers were scored. The number of flowers was limited by the relatively small sample size allowed (Fig. 5B) and the environmental stresses of the 2-day orbital flight.

In some of the earlier stamen-hair experiments, useful results were obtained by scoring all six stamens from each of three flowers per exposure per day. However, our experience suggests that more variation is likely to occur among stamens from different cuttings than among the stamens from the same flower. Thus, if only 18 stamens are scored per treatment per day, statistically better results are likely to occur if single stamens from 18 different cuttings are scored. Using 25 cuttings per exposure,[56–60] it usually was possible to have 18 flowers produced each day during only the earlier part of the scoring period. When flower production dropped, additional stamens were taken from each flower to maintain the required number per exposure. Although 18–20 stamens are often adequate at higher exposure levels, these numbers become insufficient at low radiation doses (less than 10 rads of X-rays or 0.5 rad of fast neutrons), particularly if dose-response curves for the rarer aberrations are desired. Day-by-day mutation frequency curves may also fluctuate. We have used over 2000 cuttings per exposure and scored several hundred stamens per dose per day using doses in the millirad region.[1] Since results are usually very reproducible, even larger totals may be obtained by doing replicate experiments and pooling the data from similar treatments.

The number of stamen hairs scored and their resulting standard errors used to construct dose-response curves at intermediate dose levels (about 20–90 rads of X-rays or lesser doses of fast neutrons) led us to extrapolate that a dose of 0.25 rad of X-rays would require the scoring of nearly 1 million hairs each for the controls and irradiated sample to establish the low dose point with an acceptable reliability.[1] From the data derived from these experiments, a computer program was devised whereby it is possible to estimate numbers of stamen hairs required to determine pink-event frequency (minus control) with a standard error which is a pre-determined percentage of the mean for any given dose of X-rays.

Figure 6 is a graph illustrating appropriate numbers of stamen hairs needed for resulting standard errors to fall within 1, 5, 10, and 20% of the mean values. This graph is based on data obtained during a 5-day scoring period (days 11–15 after irradiation) which encompasses the days of highest mutation frequency for pink mutations.

The left-hand scale on Fig. 6 shows total numbers of stamen hairs needed and total numbers of flowers (based on 300 hairs per flower). The right-hand scale gives the numbers of stamen hairs and flowers needed each day for a 5-day scoring period. Aberration frequency and X-ray dose are shown along the abscissa.

It is usually desirable to determine the number of flowers or stamen hairs to be scored for each day rather than the total number, since it must be decided how much scoring can be undertaken each day. The experimenter, of course, must decide on the standard errors he is willing to

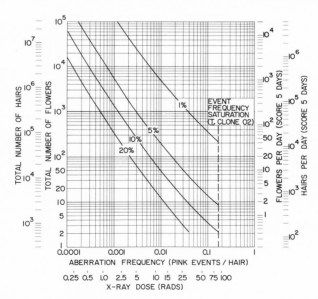

FIGURE 6. Guide for predicting population size required at various doses of X-rays to determine *Tradescantia* clone 02 pink-event frequency (minus control) with a standard error which is a predetermined percentage of the mean. Graph is based on data collected during a 5-day scoring period (days 11–15 after irradiation, assuming 300 hairs per flower).

accept, and these will, in many cases, be determined or limited by the amount of technical staff available.

According to Fig. 6, it is possible to get small standard errors by scoring only one to three flowers a day as the region of saturation is approached. In practice, this is not recommended unless individual stamens are selected from several different cuttings, because there is often variation in responses among cuttings. We suggest that if possible the experimenter score about ten flowers per day even though fewer might suffice. Using this value, one could go to doses as low as 5 rads and still have a standard error of about 20% of the mean. At 1 rad, over 100 flowers would be needed to fall within the 20% range. At still lower doses, it seems almost certain the experimenter would have to settle for standard errors in excess of 20%.

For most radiations except those having an LET lower than X-rays, the numbers of stamen hairs scored usually can be less than the numbers indicated in Fig. 6 because of the relatively high RBE values expected from this species.

During the course of an experiment, an adequate number of control stamen hairs must also be scored, preferably a number nearly equal to the number scored for the lowest exposure used. Seven separate experiments were necessary to construct the X-ray and neutron dose–response curves down to values less than 1 rad.[1] Separate controls were scored for each experiment, and the results are shown in Table 2. For each of these experiments, the frequency of spontaneous pink mutant events per stamen hair (total mutant events/total hairs taken through a 5-day scoring period), standard errors, total numbers of stamen hairs, and the total number of mutant events are listed. These control values are for the most part very low, which illustrates the relatively high stability of the system.

Figure 6 should be equally useful in determining numbers of stamen hairs to be scored using chemical mutagens if the initial concentration of the mutagen and the mutation frequency are known. The aberration frequency that resulted from the unknown gaseous mutagen which entered the growth chamber at Brookhaven[12] is shown to be approximately equal to that of 90 rads of X-rays in Fig. 12. As stated above, a 1.0 mM aqueous solution of nitrosoguanidine yielded similar results. Lesser concentrations would be expected to yield aberration or mutation frequencies equivalent to lesser doses of X-rays and thus require the scoring of more stamen hairs. Although the shapes of dose–response curves for various chemical mutagens are not yet known, they should resemble those for X-rays or neutrons sufficiently so that Fig. 6 will be useful as a scoring guide.

TABLE 2. Frequency of Spontaneous Pink Mutant Events per Hair in Stamens of *Tradescantia* Clone 02 in Seven Separate Experiments[a]

	0.43-MeV neutrons					250-kVp X-rays			
Expt. No.	Mutant events per hair	s E	Total mutant events	Total number of stamen hairs scored	Expt. No.	Mutant events per hair	s E	Total mutant events	Total number of stamen hairs scored
Control 1	0.0015	0.0008	4	2,636	Control 4	0.0035	0.0018	5	1,431
Control 2	0.0006	0.0004	2	3,395	Control 5	0.0029	0.0012	6	2,094
Control 3	0.0007	0.0001	61	87,320	Control 6	0.00087	0.00011	72	82,972
					Control[b] 7	0.00065	0.00007	622	938,408

[a]Values averaged from total mutant events divided by total stamen hairs scored from days 11-15 after irradiation.
[b]Two-part experiment in which 402,548 and 535,860 stamen hairs were scored.

G. Handling the Cuttings during Experimentation

Cuttings appear able to withstand rather harsh manipulations, but since the yield of mutations can be modified by various physical and chemical environmental conditions, the need for adequately controlled experiments and properly handled cuttings must be emphasized. [12,47,48]

It is desirable to treat the cuttings as gently as possible, but in practice this cannot always be done. Experiments with monoenergetic neutrons necessitated that we remove not only the older buds but a large part of the bracts surrounding the inflorescence as well as much of the leaf material (Fig. 5A). Judging from the controls, this particular treatment, although very harsh, did not appear to affect the mutational responses of the stamen hairs. However, it may have adversely affected stamen-hair number and/or flower production, since excessive bud blasting (killing) was found in certain instances. Bud blasting may result from exposure to low humidity for only a few hours. It is therefore essential to prevent desiccation by covering the buds with moist cheesecloth or otherwise maintaining a high humidity.

Severe bud blasting has also been shown to occur following a 2-day exposure to an environment containing as little as 0.9 ppm of ethylene. [48] Therefore, careful consideration should be given to the levels of ethylene and other environmental contaminants or possible mutagens such as SO_2 and ozone when critical mutation experiments are undertaken and there is reason to suspect the presence of contaminants. [12] As a precaution, the air intakes to our growth chambers are equipped with activated charcoal filters.

For protection from desiccation during long exposure periods, the basal end of each cutting is first wrapped in moist cotton or gauze. Then the cuttings may be inserted into tubes of very thin flexible polyethylene (4 cm diameter) previously cut to the length of the cutting excluding the inflorescence, folded over, and stapled at one end or, as shown in Fig. 5A, wrapped in Saran Wrap (thin sheet plastic). Such a package may be taped or otherwise maneuvered into the desired exposure position. These procedures were designed for specific radiation experiments and probably would not be necessary for most chemical mutagen experiments, in which exposures may be made using whole plants or cuttings in beakers.

Experiments which require an extended treatment of 2–4 days necessitate a reservoir of nutrient solution for the cuttings. In the Biosatellite II experiment, this was accomplished by inserting the base of a fresh axillary cutting through a rubber seal in the cap of a nutrient tube, rooting the cutting in vermiculite for the required 7–10 days, and then inserting the roots into the 2-ml nutrient tube [48] (Figs. 3 and 5B). Following treatment, the rooted cuttings are removed from the tube and placed in aerated nutrient solution with the cap left on the stem, since removal would damage the roots. Such a procedure might also be useful where it is desirable to expose the cuttings in a small volume, i.e., a rare or expensive mutagen.

H. Types of Aberrations Found in Stamen Hairs

Several types or combinations of somatic aberrations may be recognized and scored. The change from the normal blue color to pink is a true mutation (or deletion), since the plant is heterozygous for flower color.[36,59,65] However, the nature of the other aberrations is unclear as yet. They could result from physiological disturbances. There are basically three color types: the dominant blue, the recessive pink, and the genetically undetermined colorless. There are five morphological types: normal-size cells, giant cells, and dwarf cells, and the bent and branched hairs that arise from these individual cell types. These distinct aberrations which can be found singly or in combination in individual cells or hairs are shown diagrammatically in Fig. 7. Figure 8C,D shows an alphabetical code useful for designating the possible aberrant cell types and combinations. Various misshapen cell forms may also appear, especially in the terminal hair cells,[30] but their occurrence, although somewhat dose dependent, is too erratic to give consistent results, and we no longer score them.[47] Other aberrations, e.g., branched and bent

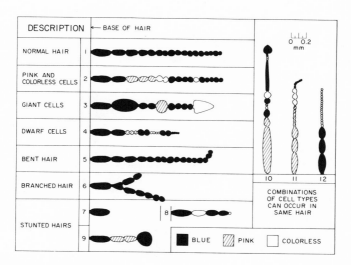

FIGURE 7. Diagram of a normal *Tradescantia* stamen hair and examples of aberrant hair and cell types observed following irradiation. (From Underbrink *et al.*[59])

A RECORDING GUIDE

B COMPUTER INPUT SHEET

EXPERIMENT X-RAY

TREATMENT X-10

DATE 6-20-71

R + 16

INPUT – OUTPUT

SCORER A.N. & C.N   PROOFED   V.P.   DATE SCORED 6-20-71   PAGE 60 OF 74

C ABERRATION CODE LETTERS

MORPHOLOGY COLOR	NORMAL	BENT HAIR	BRANCH- ING	GIANT	DWARF	GIANT BENT HAIR	DWARF BENT HAIR	GIANT BRANCH- ING	DWARF BRANCH- ING
BLUE	A	C	D	E	F	O	P	Q	X
PINK		G	H	I	J	R	S	T	Y
COLOR- LESS	B	K	L	M	N	U	V	W	Z

D COMBINATIONS OF ABERRATIONS

PINKNESS----------- AGHIJRST
COLORLESSNESS------ BKLMNJVW
DWARFISM----------- FJNPSV
GIANTISM----------- EIMOQRSUV
BENT HAIRS--------- CGKOPRSUV
BRANCHED HAIRS----- DHLQTW
TOTAL ABERRATIONS-- A-Z

EVENT SEQUENCES	NUMBER OF SEPARATE EVENTS	TYPE EVENTS	SCORING CODE	NUMBER OF COMBINED EVENTS	TYPE EVENTS
[A]	1	PINK NORMAL	A	1	PINKNESS
[A][A]	2	PINK NORMAL	A	2	PINKNESS
[A][A][A]	2	PINK NORMAL	A	2	PINKNESS
[A I J A]	4	PINK NORMAL / PINK GIANT / PINK DWARF	A / I / J	1	PINKNESS
[A A A A A A A]	1	PINK NORMAL	A	1	PINKNESS
[A B A]	3	PINK NORMAL / COLORLESS NORMAL	A / B	2 / —	PINKNESS / COLORLESSNESS
[N N B M B]	4	COLORLESS DWARF / COLORLESS NORMAL / COLORLESS GIANT	N / B / M	1	COLORLESSNESS
[E I M] [E E E]	4	BLUE GIANT / PINK GIANT / COLORLESS GIANT	E / I / M	2 / — / —	GIANTISM

E

FIGURE 8. A, Scoring guide used with computer input sheets to aid in mapping the position of somatic aberrations within stamen hairs. B, Computer input sheet. Alphabet code is used to designate the type and position of somatic aberrations as explained in text. C, Alphabet code used at Brookhaven to designate aberrant cell and hair types. Eight cell types (heavier boxes) are most commonly found. D, Combinations of code letters by which the computer can scan for various combined morphological and color aberrations. E, Examples of aberrant-event sequences. Depending on experimental design, the aberrant cells may be considered as separate events or combined into morphological and color events (see text).

hairs (Fig. 7), either occur too infrequently or have an irradiation response too erratic to produce interpretable dose-response curves. Eight common aberrant cell types or combinations of types that occur in usable frequencies are pink normal-size cells, giant and dwarf cells, colorless normal-size cells, giant and dwarf cells, and blue giant and dwarf cells (Fig. 8C). All of these aberrant cell types were also observed in plants accidentally exposed to the unknown gaseous chemical mutagen and in those treated with a chemical mutagen in water.[12] We have always observed the same spectrum of mutant and aberrant forms regardless of the quality of radiation as well. Effects from treatment to treatment appear to be strictly quantitative.

Generally, there is little difficulty in recognizing most of these aberrations. However, some borderline cases are found. Aberrant cell types have been defined in order to increase objectivity when it may be difficult to classify a cell with respect to being normal or aberrant. A list of such definitions is particularly important to obtain uniform results when several scorers are involved, and it should help ensure better uniformity in results from different laboratories.

1. Pink Cells

In clone 02, the pink cells are usually brilliant pink (Petunia Purple, Color No. 32[74]), but occasionally they may appear as deep maroon, or they can be extremely faint in hue. However, a cell is scored as pink if *any* pink pigment can be recognized (Figs. 7 and 9). A didimium filter (Fish-Schurman Corp., New Rochelle, N.Y.) can be used to intensify the pink color and improve blue-to-pink contrast.[75]

2. Colorless Cells

Colorless cells can be more difficult to score than pink cells because some which at first glimpse appear colorless will be found on more careful examination to have a frosted or translucent appearance or a very faint bluish hue. It is difficult to use a didimium filter when scoring for these cells because the filter imparts a pinkish cast to the colorless cells. Only those cells which are absolutely clear, transparent, and completely devoid of visible pigmentation are scored as colorless cells (Figs. 7 and 9).

3. Giant and Dwarf Cells

Giant and dwarf cells may be blue, pink, or colorless. They are sometimes difficult to recognize simply by comparing them with cells in the same position in adjacent normal hairs, so an *additional* criterion is that they should show an abrupt change in size compared with normal-size cells more proximal to the filament within the same hair. Beyond this, the

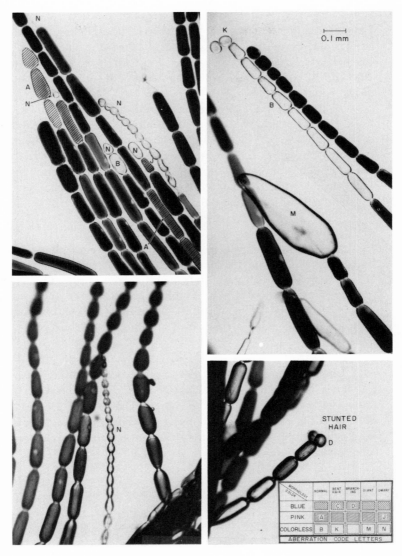

FIGURE 9. Photomicrographs showing various aberrations observed in stamen hairs following irradiation or treatment with chemical mutagens.

giant cells must be at least twice the size in either length or width of the cell in the same position in adjacent hairs. They are often misshapen as well. Occasionally, giant cells may be extremely large (Figs. 7 and 9). To be scored as a dwarf, the cell should be at most one-third the size in either length or width of any cell in the same position in adjacent normal hairs, but here too there must be an abrupt decrease in cell size compared to the other adjacent normal-size cells in the same hair. Dwarf cells often occur in long chains, but when they occur singly or as doublets or triplets they can be easily missed in scoring because of their extremely small size (Figs. 7 and 9).

There is normally a progression of diminishing cell sizes from large basal cells to much smaller cells in the more distal positions of the hair and a diminution of cell sizes in hairs from the base of the filament toward the anther (Figs. 2, 7, and 9). Therefore, it is important that giant and dwarf cells be compared with cells which are approximately in the same cell position from the filament in adjacent normal hairs and also with adjacent cells in the same hair.

4. Bent Hairs

There is an abrupt change in linearity in bent hairs, and the hair cannot be straightened under the dissecting stereoscope with a probe. This change presumably results from a shift in the normal plane of division during mitosis. The resulting daughter cell is located at an angle to the existing file of cells (Figs. 7 and 9). Sometimes the bend occurs in several contiguous cells, which results in a spiraling of the hair.

5. Branched Hairs

Branched hairs are easy to recognize but are usually rare in clone 02. They occur when two cells arise from one cell; one or both of the daughter cells may continue to divide (Figs. 7 and 9). Because of the format of the computer input sheets (described below), only aberrations and cell numbers which occur in the longer arm of each branched hair are scored.

I. Definition of a Mutant or Aberrant Event

Since the nature of the induced cell abnormalities is not clear (except for pink) we usually refer to them jointly as aberrations or aberrant events rather than mutant events. However, this is meant to be a noncommittal term and is not meant to imply that some of the events are of a nongenetic nature.

For scoring purposes, we distinguish an aberrant or mutant *event* from an aberrant or mutant *cell* as follows: If a single aberrant cell of any aberrant type is found either between normal blue cells or between some other type of aberrant cell (e.g., a pink cell between two blue cells, or a pink cell between two blue giant cells), the pink cell is termed a single mutant event. Two or more contiguous cells of the same aberrant type are also considered as a single event, since it is highly probable that only one change occurred during treatment when the hairs were immature and, upon division, the original aberration was carried to daughter cells. When the entire hair consists of a single aberrant cell type, e.g., all contiguous pink normal-size cells, it is still termed a single event. Should a hair have two pink cells separated by a blue or colorless cell or cells, it is scored as having two pink events, plus the number of events resulting from the presence of other aberrant cell types within the same hair. The same method is used for all types of aberrations. Thus, the number of aberrant or mutant *cells* is always higher than the number of aberrant *events* except for the first few days of posttreatment scoring, when it may be almost the same. [59]

Aberrant cells containing double phenotypic changes, e.g., colorless dwarf or pink giant, are scored with the appropriate symbol on the computer input sheet but, if desired, may be totaled as one single type of event with respect to either cell size or color in plotting dose-response curves. Pink giant cells can thus be grouped as a single event for "pinkness" with contiguous pink dwarf cells, or they might be grouped as a single event for "giantism" with contiguous blue giant cells.

Such cell types more probably represent changes in two different genetic loci. If these are on the same chromosome, a single deletion could account for them, but if on different chromosomes they must represent two different and distinct genetic changes. One may choose to ignore the latter possibility and regard them, partly for convenience, as single events of either cell type. Similar examples of scoring are illustrated in Fig. 8E.

J. Loss of Reproductive Integrity (Stamen-Hair Stunting)

Another change brought about by suitable doses of radiation is a reduction in cell number and hence stamen-hair length. This is a loss of reproductive integrity and is referred to as "hair stunting" (Figs. 7 and 9). Since each hair appears to develop from a one- or two-celled meristem, [15,21,27,29,30] they, in a sense, are analogous to a small cell colony. Thus, if one determines reproductive incapacitation, a surviving hair fraction can be calculated and typical survival curves can be made.

For the determination of hair stunting, the number of cells per hair is required. Hair stunting can occur in all hairs throughout the stamen. However, the cell number per hair in the upper one-third of the hair-bearing region is somewhat less than in the lower two-thirds, so it is more convenient to use only hairs from the upper region. In fact, because the cell number per hair decreases slightly from base to apex of the filament, we have found that for scoring purposes, it is convenient to divide (by eye) the hair-bearing region of the stamen into three equal sections (Fig. 2B).

The mean hair length in the upper region is about 16 cells per hair,[41,47,59] while hairs in the middle and lower regions can be longer. [41,66] In control populations, 95% of the hairs in region 3 consist of 12 or more cells. Therefore, we routinely consider irradiated hairs 12 or more cells in length as "survivors," while those consisting of fewer cells are stunted ("nonsurvivors"). However, in practice, survival curves can be made with hairs of various lengths considered as survivors.[59] Following irradiation, stunting in region 3 reaches a maximum on days 17–19 after irradiation.[55]

Data for hair stunting from treatments other than ionizing radiation are not yet available.

K.   Preparation of Materials for Scoring

To begin the scoring procedure, a stamen is carefully removed at its base with a fine forceps. Since stamen hairs occur all the way to the base of the filament, care should be exercised to ensure a minimum of damage to the basal hairs. One may gently blow on it to remove loose pollen. The stamen is placed in a small drop of paraffin oil (Saybolt Viscosity 125/135) on a glass slide with the adaxial side of the filament down. Hairs are not normally formed on this surface, and hence few will be hidden by the filament.

Using fine dissecting probes under a stereoscope (about 25 × magnification), the hairs can be straightened to lie flat on both sides of the filament (Fig. 2B). The correct amount of paraffin oil may be somewhat difficult to ascertain at first. If the oil is insufficient, the hairs may tend to stick to the slide, break, or be difficult to maneuver. On the other hand, if excessive oil is used, the hairs tend to float and move about freely and may become more difficult to examine in one focal plane.

At room temperature, the hair cells may begin to plasmolyze within a few minutes after placing the stamen on the dissecting stereoscope stage. Plasmolyzed cells are difficult to score reliably, and some cannot be scored at all. Thus, the hairs should be scored as quickly as possible, but speed in scoring requires practice. Some scorers prefer to retard plasmolysis by placing small pieces of dry ice under the stage of the stereoscope. Thus cooled, the stamen hairs last much longer. A water filter may be used in the light beam to decrease heat. This also retards plasmolysis. Stamens in a drop of paraffin oil on a slide may also be placed directly on dry ice and stored frozen for subsequent interpretation or photographic purposes.

L. Scoring Procedures

We have found it useful to use three different scoring procedures depending on the nature of the experiment and its objectives.

1. Method I

If the experiment does not require detailed information, a simple scanning of the stamen for the number of pink events may suffice. This is very fast and simple and may be accomplished by a single scorer who tabulates the number of pink events in the entire stamen. No distinction is made between aberrant cells of different sizes. Thus, a contiguous pink normal-size cell, pink giant, and pink dwarf would be scored as one single event. The day before scoring and the day after scoring and perhaps 1 day during scoring (assuming a 5-day scoring period), the average number of hairs per stamen is determined. This number varies but is on the order of 50 hairs per stamen (300 per flower) under our growing conditions. Data are calculated as number of events per hair rather than per cell. For this method, input sheets can be simplified and a much less complicated computer program can be used or omitted completely. Although limited, this method is extremely valuable, and its ease has made it possible for us to score the vast numbers of stamen hairs needed for experiments at very low dose levels.[1] This type of scoring should prove valuable for screening possible chemical mutagens or arriving at proper chemical concentrations to use in more elaborate experiments.

2. Method II

Method II is similar to method I but is used for experiments in which it is desirable to determine the frequencies of other types of events in addition to pink. For this type of scoring, a single scorer tabulates (using a multiple cell counter) the desired types of various aberrant events. All types of events may be tabulated as either individual events or groups of events. For example, it might be desirable to keep separate tabulations for blue giants, colorless giants, and pink giants. If these cell types are contiguous, he may score these as three events, or if interested only in "giantism" he may score these as a single event. Hair counts are made as in method I. Method II, of course, does not show the position of these events within the hair.

3. Method III

Method III is more difficult and complicated than the previous methods, but it supplies the most detailed information on hair length and the kinds and positions of aberrant cells within a hair as well as a record of daily changes that occur during the course of the scoring period. This scoring may proceed from day 7 after irradiation, continue through the usual aberration peaks (days 11–15), and on through the period of maximum stunting (days 17–19) or even beyond. The experimenter may "map" and describe completely all cells in each hair. This method is used to establish survival curves. It is most convenient to limit this type of scoring to the upper one-third of the filament (region 3, Fig. 2B).

Two scorers are recommended for this scoring procedure. One person scores and dictates to the second member of the team, who enters the information directly on the input sheets. After scoring about ten stamens, they alternate these assignments to reduce fatigue.

An example of a typical input sheet as used for this method is shown in Fig. 8B, but this sheet may be modified to accommodate appropriate data if method I or II is used. The example shown here was designed to accommodate two hairs on a single line, and each hair could have a maximum length of 28 cells. In this example, the experiment was done with X-rays, the "X-10" designating the exposure given. "R + __ " indicates the day after treatment, i.e., 16 days after irradiation.

A ruler or guide (Fig. 8A) is made by taping a piece of input sheet to a thin piece of plastic and numbering the spaces as shown. On the left-hand side of the guide, there are spaces which indicate the proper placement of the following formation on the input sheets: plant cutting number, flower number, day after irradiation, stamen number (Nos. 1–6, Fig. 2A), position or region of hairs on the stamen being scored (Nos. 1, 2, or 3, Fig. 2B), and the number of stamen hairs in that position. Flower number is entered on the sheets from the flower production records after the day's scoring is completed. In the earlier Biosatellite experiments, the length of the hair was denoted by an "X" (representing the filament) placed in a box one number larger than the length of the hair.[48] The current procedure is to write the number of cells for each hair to the right of the hair description as shown in Fig. 8B. A code using letters of the alphabet to designate individual aberrations is placed in the appropriate box corresponding to their actual position within the hair. The position below box No. 1 on the ruler scoring guide corresponds to the terminal or distal hair cell. The first line in Fig. 8B describes an 11-celled hair having an "A" aberration in box No. 1 or a single pink, normal-size cell in the terminal position.

Once the stamen hairs are mapped in this fashion, computer programs can be devised to analyze the data as required by the experimenter.

III. CHARACTERISTIC DATA OBTAINED FROM THE STAMEN-HAIR SYSTEM

When experiments are performed with clone 02, there are certain basic types of information desired. This information concerns aberration or mutation frequencies, dose-response curves, relative biological effectiveness (RBE) of one treatment or radiation compared to another, survival curves, mutation rates, and relationships of specific aberrant cell types to each other, to the cell number in the hair, and to their positions within the hair.

Our data are usually analyzed by the computer, and the results are returned on a "per hair" or "per cell" basis. Results may also be returned on a day-by-day posttreatment basis as well as summarized over any chosen period of scoring.

Figure 10 shows typical day-to-day responses obtained from the total combined aberrations. It is readily seen that the aberrant events are dose dependent and that their peak values are reached during a scoring period encompassing days 7 through 20 after irradiation (for "pinkness," peak days in rooted cuttings usually appear to be days 11–15 after irradiation[1]). However, it is often difficult to determine the true peak day, since there may be large daily fluctuations unless very large samples are scored. Smooth curves may be mathematically fitted through the

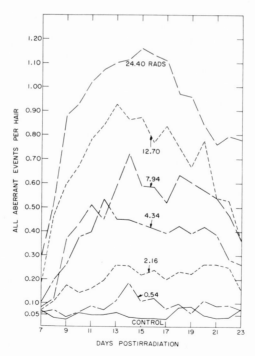

FIGURE 10.   Day-to-day aberration response frequencies obtained by scoring for and combining all aberrant cell types following a series of doses with 0.43-MeV neutrons.

data points, giving computed peak values.[48] This is advantageous in instances when it is impractical to score very large numbers of stamen hairs. Should numbers of stamen hairs be insufficient to determine precisely the exact peak day for aberration frequency, another approach is to take mean values around the presumed peak day, i.e., 2 days on either side. Still another approach is to use cumulative aberration frequencies. Cumulative values (Fig. 11) are obtained by the addition of each day's new data to that of the preceding day, thus diluting the apparent day-to-day variation. RBE values can be obtained from dose-response curves constructed from maximum aberration frequencies provided the peak day can be determined accurately. In general, however, the shapes and slopes of the dose-response curves (and thus RBE values) are not appreciably different when either peak aberration values or cumulative values are used (Fig. 12). Cumulative scoring is most useful when no scoring days are missed. This requires assistants to work 7 days a week. Cumulative data are also harder to interpret in terms of mutation rates. For these reasons, we now routinely collect data for only the 5-day peak period. It is necessary, of course, to be reasonably sure that the 5 days chosen are really those encompassing the true aberration peak.

Useful data are obtained between days 7 and 20 after treatment. Prior to day 7 after treatment, the hairs were too mature to respond to treatment (i.e., no more cell divisions), and after day 21 few of the hairs were present at time of treatment so that responses are relatively low.

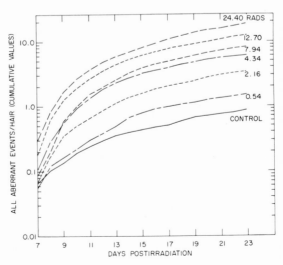

FIGURE 11. Cumulative aberration responses obtained by scoring for and combining all aberrant cell types for each day after treatment following irradiation with 0.43-MeV neutrons.

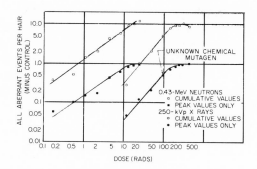

FIGURE 12. Dose-response curves obtained after 0.43-MeV neutron and X-irradiation by scoring for and combining all aberrant cell types. Cumulative frequencies (days 7–20 after irradiation) are higher than peak frequencies, but the slopes and thus RBE values do not change appreciably. The unknown chemical mutagen that entered a growth chamber at Brookhaven[12] yielded an aberration frequency approximately equal to 90 rads of X-rays.

It is possible to distinguish three types of stamen-hair populations for studies involving hair stunting. The first is simply the entire hair population consisting of both normal and abnormal hairs, the second is that population of hairs which contains no aberrant cells, and the third is that population which consists only of those hairs with particular types or combinations of aberrations. Survival curves may be obtained from each population.[55] Figure 13 shows examples of survival curves which

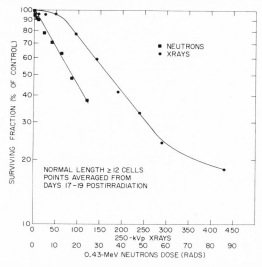

FIGURE 13. Survival curves for stamen hairs after neutron and X-irradiation. The surviving fraction is the percent of normal-length hairs as percent of control for the terminal third of the filament. The surviving fraction consists of hairs both with and without aberrations.

include both normal and visibly aberrant hairs from region 3 of the filament (after irradiation with 0.43-MeV neutrons and 250-kVp X-rays) in which hairs consisting of 12 or more cells are considered as survivors. By separating normal and aberrant stamen hairs into separate populations, it is possible to determine the extent to which loss of reproductive capacity occurs in hairs having no visible aberrations.

Suitable computer programs also enable the study of the number of cells per hair preceding the first aberrant event. Such studies may be able to provide information on the early development of the stamen hairs. For example, in the early days after irradiation (R + 7–10), many of the hairs in region 3 were 14–15 cells in length at the time of treatment, as evidenced by the occurrence of aberrant events at the tips of hairs 14–15 cells long. The mean length of the hairs proximal to the cell in which the first aberrant event occurs then continues to shorten as the time after treatment lengthens until about day 17 after treatment, when most aberrant events are in the basal hair cells. This indicates that at the time of treatment either filament epidermal cells or perhaps only single-celled hairs were present. It should be noted that this approach to "aberration dating" must take into consideration the fact that from 7 to 15% of the cell divisions in stamen hairs are in interstitial cells and not solely in apical or subapical cells. [15,27,41]

Thus, handled properly, the stamen hair system of *Tradescantia* clone 02 can provide some unique mutagen information. Besides the resemblance of the growing stamen hair to a small cell culture colony, it has further advantages in that visible aberrant or mutant cell types are produced and cell lineage is clearly defined. In addition, the system is easily handled so that statistically adequate populations are usually readily obtained. Its advantage over a bacterial system is that it has eukaryotic chromosomes and is very sensitive to physical and chemical mutagens.

Although clone 02 appears to be the most suitable and versatile of the commelinaceous plants for mutagenic studies, new and improved clones are being or have been developed. [31]

Again, it should be pointed out that the techniques described above were developed for radiation experiments, but since *Tradescantia* stamen hairs are affected by at least some chemical mutagens, these techniques should be for the most part applicable to chemical mutagenesis.

IV. SUMMARY

Detailed procedures for the proper and efficient handling of the *Tradescantia* clone 02 stamen-hair system are given. The techniques described were developed primarily to utilize stamen hairs as a radiobiological test system, but preliminary experiments indicate that the system and techniques are also applicable to chemical mutagenesis.

Specific areas discussed include advantages and uses of the stamen-hair system, review of floral and stamen-hair morphology, cultivation of stock plants, screening of untreated plants for abnormally high aberration rates, preparation and handling of cuttings for experimental use, and consideration of flower production and of statistically acceptable numbers of cuttings and stamens required for a typical experiment.

Although there are many advantages offered by the stamen-hair system, it is of particular interest because it has a eukaryotic nucleus, is heterozygous for blue–pink flower color genes, and may be employed in numbers sufficient to detect very low levels of radiation (and presumably of chemical mutagens) that might not be practical with many other types of organisms. Consideration of numbers of stamen hairs scored in several previous radiation experiments at several dose levels and the standard deviations obtained made it possible to calculate numbers of stamen hairs theoretically needed to come within desired limits of error down to very low doses.

Descriptions of the types and combinations of visible somatic mutations or aberrations found in stamen hairs and details of the computer-oriented scoring techniques are given. Several types of data characteristically obtained from the stamen-hair system are discussed. These include cell mutation frequencies and their dose-response curves, relative biological effectiveness, cell survival curves, mutation rates, and relationships of specific aberrant cell types to each other, to the cell number in the hair, and to their positions within the hair.

V. ACKNOWLEDGMENTS

In compiling this report, the authors recognize and appreciate that the following scientists were responsible for developing certain aspects of the stamen-hair techniques now in use: Dr. R. L. Cuany, Dr. D. R. Davies, Dr. S. Ichikawa, Dr. K. M. Marimuthu, Dr. L. W. Mericle, Dr. Rae P. Mericle, and Dr. G. G. Nayar. The authors are particularly indebted to Mr. K. H. Thompson, who has written the necessary computer programs used at Brookhaven National Laboratory without which analysis of the stamen-hair data would be severely limited. The authors are also grateful to the following technicians at Brookhaven who contributed substantially to the development of certain of the techniques: Mrs. Priscilla M. Baetcke, Mrs. Marta M. Nawrocky, Mr. R. C. Sautkulis,

Mrs. Marie U. Schairer, and Mrs. Rhoda C. Sparrow. The authors thank Miss Virginia Pond for suggestions concerning the manuscript.

VI. REFERENCES

1. A. H. Sparrow, A. G. Underbrink, and H. H. Rossi, Mutations induced in *Tradescantia* by small doses of x-rays and neutrons: Analysis of dose-response curves, *Science 176*, 916–918 (1972).
2. H. H. Smith and T. A. Lotfy, Comparative effects of certain chemicals on *Tradescantia* chromosomes as observed at pollen tube mitosis, *Am. J. Botany 41*, 589–593 (1954).
3. P. S. Rushton, The effects of 5-fluorodeoxyuridine on radiation-induced chromatid aberrations in *Tradescantia* microspores, *Radiation Res. 38*, 404–413 (1969).
4. L. Ehrenberg, in "Chemical Mutagens: Principles and Methods for Their Detection" (A. Hollaender, ed.), Vol. 2, pp. 365–386, Plenum Press, New York (1971).
5. L. Fishbein, W. G. Flamm, and H. L. Falk, "Chemical Mutagens," Academic Press, New York (1970).
6. K. Gröber, F. Scholz, and M. Zacharias (eds.), "Induzierte Mutationen und ihre Nutzung: Erwin-Baur-Gedächtnisvorlesungen IV, 1966," Akademie-Verlag, Berlin (1967).
7. B. A. Kihlman, "Action of Chemicals on Dividing Cells," Prentice-Hall, Englewood Cliffs, N.J. (1966).
8. B. A. Kihlman, in "Chemical Mutagens: Principles and Methods for Their Detection" (A. Hollaender, ed.), Vol. 2, pp. 489–514, Plenum Press, New York (1971).
9. A. Loveless, "Genetic and Allied Effects of Alkylating Agents," Butterworths, London (1966).
10. "Symposium on Chromosome Breakage," *Heredity 6* (Suppl.) (1953).
11. A. H. Sparrow, P. J. Bottino, and L. A. Schairer, in "Report of the Secretary's Commission on Pesticides and Their Relationship to Environmental Health," Parts I and II, pp. 575–584, U.S. Department Health, Education and Welfare, Washington, D.C. (1969).
12. A. H. Sparrow and L. A. Schairer, Mutational response in *Tradescantia* after an accidental exposure to a chemical mutagen, *EMS Newsletter* No. 5, 16–19 (1971).
13. K. A. Jensen, I. Kirk, G. Kølmark, and M. Westergaard, Chemically induced mutations in *Neurospora, Cold Spring Harbor Symp. Quant. Biol. 16*, 245–261 (1951).
14. E. Cerdá-Olmedo and P. C. Hanawalt, Diazomethane as the active agent in nitrosoguanidine mutagenesis and lethality, *Mol. Gen. Genet. 101*, 191–202 (1968).
15. A. H. Sparrow and L. A. Schairer, Unpublished data.
16. K. Sax, Chromosome aberrations induced by X-rays, *Genetics 23*, 494–516 (1938).
17. M. R. Alvarez and A. H. Sparrow, Comparison of reproductive integrity in the stamen hair and root meristem of *Tradescantia paludosa* following acute gamma irradiation, *Radiation Botany 5*, 423–430 (1965).
18. K. P. Baetcke, C. H. Nauman, A. H. Sparrow, and S. S. Schwemmer, The relationship of DNA content to nuclear and chromosome volumes and to radiosensitivity (LD$_{50}$), *Proc. Natl. Acad. Sci. 58*, 533–540 (1967).
19. A. D. Conger and N. H. Giles, The cytogenetic effect of slow neutrons, *Genetics 35*, 397–419 (1950).
20. A. D. Conger, M. L. Randolph, C. W. Sheppard, and H. J. Luippold, Quanti-

tative relation of RBE in *Tradescantia* and average LET of gamma-rays, x-rays, and 1.3-, 2.5-, and 14.1-MeV fast neutrons, *Radiation Res. 9,* 525–547 (1958).

21. D. R. Davies, Radiation-induced chromosome aberrations and loss of reproductive integrity in *Tradescantia, Radiation Res. 20,* 726–740 (1963).

22. D. R. Davies and J. L. Bateman, A high relative biological efficiency of 650-keV neutrons and 250-kVp x-rays in somatic mutation induction, *Nature 200,* 485–486 (1963).

23. D. R. Davies, A. H. Sparrow, R. G. Woodley, and A. Maschke, Relative biological efficiency of negative μ-mesons and cobalt-60 γ-rays, *Nature 200,* 277–278 (1963).

24. J. E. Gunckel, A. H. Sparrow, I. B. Morrow, and E. Christensen, Vegetative and floral morphology of irradiated and non-irradiated plants of *Tradescantia paludosa, Am. J. Botany 40,* 317–332 (1953).

25. S. Ichikawa, A radiobiological study in the stamen hairs of *Tradescantia blossfeldiana* Mildbr., *Seiken Ƶihô 20,* 35–45 (1968).

26. S. Ichikawa, RBE of 14.1 MeV fast neutrons and ^{137}Cs gamma rays in the stamen hairs of *Tradescantia reflexa, Genetics 64* (Suppl. 2), s31 (1970).

27. S. Ichikawa and A. H. Sparrow, Radiation-induced loss of reproductive integrity in the stamen hairs of a polyploid series of *Tradescantia* species, *Radiation Botany 7,* 429–441 (1967).

28. S. Ichikawa and A. H. Sparrow, Radiation-induced loss of reproductive integrity in the stamen hairs of *Tradescantia blossfeldiana* Mildbr., a twelve-ploid species, *Radiation Botany 7,* 333–345 (1967).

29. S. Ichikawa and A. H. Sparrow, The use of induced somatic mutations to study cell division rates in irradiated stamen hairs of *Tradescantia virginiana* L., *Japan. J. Genet. 43,* 57–63 (1968).

30. S. Ichikawa, A. H. Sparrow, and K. H. Thompson, Morphologically abnormal cells, somatic mutations and loss of reproductive integrity in irradiated *Tradescantia* stamen hairs, *Radiation Botany 9,* 195–211 (1969).

31. A. Kappas, A. H. Sparrow, and M. M. Nawrocky, Relative biological effectiveness (RBE) of 0.43-MeV neutrons and 250-kVp x-rays for somatic aberrations in *Tradescantia subacaulis* Bush, *Radiation Botany 12,* 271–281 (1972).

32. K. M. Marimuthu, L. A. Schairer, and A. H. Sparrow, The effects of space flight factors and gamma radiation on flower production and microspore division and development in *Tradescantia, Radiation Botany 10,* 249–259 (1970).

33. K. M. Marimuthu, L. A. Schairer, A. H. Sparrow, and M. M. Nawrocky, Effects of space flight (Biosatellite II) and radiation on female gametophyte development in *Tradescantia, Am. J. Botany 59,* 359–366 (1972).

34. K. M. Marimuthu, A. H. Sparrow, and L. A. Schairer, The cytological effects of space flight factors, vibration, clinostat and radiation on root tip cells of *Tradescantia, Radiation Res. 42,* 105–119 (1970).

35. L. W. Mericle and R. P. Mericle, Biological discrimination of differences in natural background radiation level, *Radiation Botany 5,* 475–492 (1965).

36. L. W. Mericle and R. P. Mericle, Genetic nature of somatic mutations for flower color in *Tradescantia,* clone 02, *Radiation Botany 7,* 449–464 (1967).

37. L. W. Mericle and R. P. Mericle, Mechanisms of somatic "mutation" induction in flowers of hybrid *Tradescantia* (clone 02), *Genetics 56,* 576–577 (1967).

38. L. W. Mericle and R. P. Mericle, *in* "Induzierte Mutationen und ihre Nutzung: Erwin-Baur-Gedächtnisvorlesungen IV, 1966" (K. Gröber, F. Scholz, and M. Zacharias, eds.), pp. 65–77, Akademie-Verlag, Berlin (1967).

39. L. W. Mericle and R. P. Mericle, *in* "Induced Mutations in Plants," pp. 591–601, International Atomic Energy Agency, Vienna (1969).

40. G. G. Nayar, K. P. George, and A. R. Gopal-Ayengar, On the biological effects of high background radioactivity: Studies on *Tradescantia* grown in radioactive monazite sand, *Radiation Botany 10*, 287–292 (1970).

41. G. G. Nayar and A. H. Sparrow, Radiation-induced somatic mutations and the loss of reproductive integrity in *Tradescantia* stamen hairs, *Radiation Botany 7*, 257–267 (1967).

42. G. J. Neary and J. R. K. Savage, Chromosome aberrations and the theory of RBE. II. Evidence from track-segment experiments with protons and alpha particles, *Internat. J. Radiation Biol. 11*, 209–223 (1966).

43. J. R. K. Savage, Demonstrating cell division with *Tradescantia*, *School Sci. Rev. 48*, 771–782 (1967).

44. L. A. Schairer, A. H. Sparrow, and K. M. Marimuthu, *in* "Life Sciences and Space Research" (F. G. Favorite and W. Vishniac, eds.), Vol. 13, pp. 19–24, North-Holland Publishing Co., Amsterdam (1970).

45. A. H. Sparrow, Comparisons of the tolerances of higher plant species to acute and chronic exposures of ionizing radiation, *Japan. J. Genet. 40* (Suppl.), 12–37 (1965).

46. A. H. Sparrow, K. P. Baetcke, D. L. Shaver, and V. Pond, The relationship of mutation rate per roentgen to DNA content per chromosome and to interphase chromosome volume, *Genetics 59*, 65–78 (1968).

47. A. H. Sparrow, L. A. Schairer, and K. M. Marimuthu, Genetic and cytologic studies of *Tradescantia* irradiated during orbital flight, *BioScience 18*, 582–590 (1968).

48. A. H. Sparrow, L. A. Schairer, and K. M. Marimuthu, *in* "The Experiments of Biosatellite II" (J. F. Saunders, ed.), NASA SP-204, pp. 99–122, NASA, Washington, D.C. (1971).

49. A. H. Sparrow, L. A. Schairer, M. M. Nawrocky, and R. C. Sautkulis, Effects of low temperature and low level chronic gamma radiation on somatic mutation rates in *Tradescantia*, *Radiation Res. 47*, 273–274 (1971).

50. A. H. Sparrow, A. G. Underbrink, and R. C. Sparrow, Chromosomes and cellular radiosensitivity. I. The relationship of D_0 to chromosome volume and complexity in seventy-nine different organisms, *Radiation Res. 32*, 915–945 (1967).

51. R. C. Sparrow, A. G. Underbrink, and K. H. Thompson, *in* "Book of Abstracts," p. 205, Fourth Internat. Congr. Radiation Res., Evian (1970).

52. D. Steffensen, Effects of various cation imbalances on the frequency of X-ray-induced chromosomal aberrations in *Tradescantia*, *Genetics 42*, 239–252 (1957).

53. A. G. Underbrink, Monoenergetic neutron experiments with *Tradescantia* inflorescences, *in* "Annual Report on Research Project," NYO-2740-6, pp. 255–268 (1969).

54. A. G. Underbrink and A. H. Sparrow, Power relations as an expression of relative biological effectiveness (RBE) in *Tradescantia* stamen hairs, *Radiation Res. 46*, 580–587 (1971).

55. A. G. Underbrink, R. C. Sparrow, and A. H. Sparrow, Relations between phenotypic aberrations and loss of reproductive integrity in *Tradescantia* stamen hairs, *Radiation Botany 11*, 473–481 (1971).

56. A. G. Underbrink, R. C. Sparrow, A. H. Sparrow, and H. H. Rossi, Preliminary report on monoenergetic neutron experiments with *Tradescantia*, *Radiation Res. 39*, 463 (1969).

57. A. G. Underbrink, R. C. Sparrow, A. H. Sparrow, and H. H. Rossi, *in* "Symposium on Neutrons in Radiobiology," pp. 373–388, U.S. Atomic Energy Commission, Div. Tech. Information, Oak Ridge, Tenn., CONF-691106 (1969).

58. A. G. Underbrink, R. C. Sparrow, A. H. Sparrow, and H. H. Rossi, RBEs of X-

rays, 0.43-MeV and 80-keV neutrons on somatic mutations and loss of reproductive integrity in *Tradescantia* stamen hairs, *Radiation Res. 43*, 246 (1970).

59. A. G. Underbrink, R. C. Sparrow, A. H. Sparrow, and H. H. Rossi, Relative biological effectiveness of x-rays and 0.43-MeV monoenergetic neutrons on somatic mutations and loss of reproductive integrity in *Tradescantia* stamen hairs, *Radiation Res. 44*, 187–203 (1970).

60. A. G. Underbrink, R. C. Sparrow, A. H. Sparrow, and H. H. Rossi, Relative biological effectiveness of 0.43-MeV and lower energy neutrons on somatic aberrations and hair-length in *Tradescantia* stamen hairs, *Internat. J. Radiation Biol. 19*, 215–228 (1971).

61. J. Van't Hof and A. H. Sparrow, Radiation effects on the growth rate and cell population kinetics of actively growing and dormant roots of *Tradescantia paludosa*, *J. Cell Biol. 26*, 187–199 (1965).

62. A. H. Sparrow and H. J. Evans, Nuclear factors affecting radiosensitivity. I. The influence of nuclear size and structure, chromosome complement and DNA content, *Brookhaven Symp. Biol. 14*, 76–100 (1961).

63. J. R. K. Savage and M. A. Pritchard, *Campelia zanonia* (L.) H.B.K.: A new material for the study of radiation-induced chromosomal aberrations, *Radiation Botany 9*, 133–139 (1969).

64. L. G. Parchman, The morphogenesis of stamen hairs of *Tradescantia paludosa*, Ph. D. thesis, Emory University, Atlanta, Ga. (1964).

65. A. H. Sparrow, R. L. Cuany, J. P. Miksche, and L. A. Schairer, Some factors affecting the responses of plants to acute and chronic radiation exposures, *Radiation Botany 1*, 10–34 (1961),.

66. R. C. Sparrow, Unpublished data.

67. K. Hummel and K. Staesche, *in* "Encyclopedia of Plant Anatomy," 2nd ed., Vol. 4, Part 5, pp. 207–271, Borntraeger, Berlin (1962).

68. J. C. Th. Uphof, *in* "Encyclopedia of Plant Anatomy," 2nd ed., Vol. 4, Part 5, pp. 1–206, Borntraeger, Berlin (1962).

69. A. G. Underbrink, A. H. Sparrow, and V. Pond, Chromosomes and cellular radiosensitivity. II. Use of interrelationships among chromosome volume, nucleotide content and D_0 of 120 diverse organisms in predicting radiosensitivity, *Radiation Botany 8*, 205–237 (1968).

70. H. H. Rossi, Energy distribution in the absorption of radiation, *Advan. Biol. Med. Phys. 11*, 27–85 (1967).

71. D. Steffensen, Induction of chromosome breakage at meiosis by a magnesium deficiency in *Tradescantia*, *Proc. Natl. Acad. Sci. 39*, 613–620 (1953).

72. K. D. Wuu and W. F. Grant, Morphological and somatic chromosomal aberrations induced by pesticides in barley, *Can. J. Genet. Cytol. 8*, 481–501 (1966).

73. A. D. Conger, A simple liquid-culture method of growing plants, *Proc. Fla. State Hort. Soc. 77*, 536–537 (1964).

74. "Horticulture Colour Chart," Vol. 1, British Colour Council in collaboration with the Royal Horticulture Society (1938).

75. R. F. Smith, Personal communication.

CHAPTER 31

Detection of Genetically Active Chemicals Using Various Yeast Systems

Friedrich K. Zimmermann*

Department of Biology
Brooklyn College of The City University of New York
Brooklyn, New York

I. BRIEF DESCRIPTION OF THE YEASTS USED IN GENETIC RESEARCH

Most work in yeast genetics has been performed with two species, *Saccharomyces cerevisiae* and *Schizosaccharomyces pombe*. For a detailed description of the respective life cycles, see Mortimer and Manney[1] in Volume 1 of this series. It must be pointed out that both organisms can be cultivated as haploids and, in the case of *Sacch. cerevisiae,* in stable diploid forms as well. Since there are a great variety of yeasts of the *Saccharomyces* type which do not behave in this ideal way, all genetic experiments should be performed with strains currently used by yeast geneticists. Such strains are physiologically dioecious (see Esser[2] for definition of this term), whereas many strains isolated from nature are self-compatible and haploid cells fuse uncontrollably to form diploids. Typically, *Saccharomyces* strains

---

* Present address: Fachbereich Biologie (10), Technische Hochschule, 61 Darmstadt, West Germany.

which are physiologically dioecious are available in a stable haploid or stable diploid form. The haploid cells can be mated to other haploid cells to form diploid zygotes which initiate a stable diploid phase. There is only one way to convert diploid cells into haploid cells, and this is via meiosis, which can be induced on special media. The advantage of *Sacch. cerevisiae* is based on the fact that each haploid cell can act as a gamete and each diploid cell can be induced to undergo meiosis, a unique situation indeed. *Schizosaccharomyces* can be cultured in a diploid state as well, but this diploid phase is not as stable as that of *Sacch. cerevisiae;* haploidization can be achieved by treatment with parafluorophenylalanine. [3,4]

Due to the wide spread use of *Sacch. cerevisiae* and the author's familiarity with this organism, this chapter will largely deal with this yeast rather than with *Schiz. pombe.*

II. TECHNIQUES

A. Media

All fungi are heterotrophs and require an organic carbon source for growth. Variable numbers and quantities of vitamins are required as well; otherwise, an inorganic salt medium supplemented with trace elements is enough to support good growth. In addition to these synthetic media, there are complex media based on yeast extract and peptone. Media recipes for *Schiz. pombe* have been described. [5,7] Media recipes for work with *Sacch. cerevisiae* will be described in detail.

Synthetic media can be prepared in two ways. The most convenient, although expensive, way is to use Bacto yeast nitrogen base without amino acids as supplied by Difco Laboratories (Detroit, Mich., No. 0919–15). However, a medium of equivalent quality can be composed according to the following formula, beginning with a list of ingredients for a stock solution:

$(NH_4)_2SO_4$	10 g
KH_2PO_4	8.75 g
K_2HPO_4	1.25 g
$MgSO_4 \cdot 7 H_2O$	5.00 g
NaCl	1.00 g
H_3BO_3	0.1 ml 0.1% soln.
$CuSO_4 \cdot 5 H_2O$	0.1 ml 0.1% soln.
KI	0.1 ml 0.1% soln.
$FeCl_3 \cdot 6 H_2O$	0.1 ml 0.5% soln.
$ZnSO_4 \cdot 7 H_2O$	0.1 ml 0.7% soln.

All these ingredients are dissolved in a final volume of 1 liter. This solution is then used as the stock solution, 100 ml of which is included in 1 liter of the final medium. It is also possible to make up a five-times concentrated stock solution.

Also included in the final medium are $CaCl_2 \cdot 2H_2O$, 1 ml from a 10.00 g/100 ml stock solution, and 1 ml of a vitamin solution of the following composition:

Biotin	0.2 mg
Thiamine	40.0 mg
Pyridoxine	40.0 mg
Inositol	200.0 mg
Ca-pantothenate	40.0 mg

This mixture is dissolved in 100 ml water.

These ingredients make a synthetic minimal medium. A synthetic complete medium can be supplemented according to the nutritional requirements expressed by the strains commonly used in a given laboratory. The following concentrations, expressed as per liter medium, are common:

Adenine sulfate	20 mg
L-Arginine–HCl	10 mg
L-Aspartic acid	10 mg
L-Glutamic acid	100 mg
L-Histidine–HCl	10 mg
L-Isoleucine	60 mg
L-Leucine	60 mg
L-Lysine–HCl	10 mg
L-Methionine	10 mg
L-Phenylalanine	50 mg
L-Serine	20 mg
L-Threonine	200 mg
L-Tryptophan	10 mg
L-Tyrosine	30 mg
L-Valine	30 mg
Uracil	10 mg

Stock solutions of these constituents can be prepared at 1% concentration level, and solution can be facilitated by adding KOH. Only aspartic acid and threonine have to be filtered, sterilized, and added after autoclaving. The other ingredients can be added before. Stock solutions can be kept sterile by adding a drop of chloroform to the bottles.

The complete medium used for *Sacch. cerevisiae* is YEP: Bacto peptone, 2.0% (Difco No. 0118–01), yeast extract, 1.0% (Difco No. 0127–01).

The best carbon source for yeast is glucose, which is usually administered at a 2% concentration.

Solid media are obtained by adding agar at 1.5% (Difco Bacto agar, No. 0140–01). This agar contains some adenine, which allows for some residual growth of adenine-requiring mutants. If this interferes with experiments (normally it does not), more expensive but pure preparations can be obtained.

It should be emphasized here that certain European manufacturers provide very good preparations of agar, yeast extract, and peptone. A competent yeast geneticist should be consulted, however, because not all agar preparations are pure enough for synthetic media.

It might appear to be a sacrilege, but the author has followed the sloppy procedure of autoclaving agar, sugar, mineral salts, vitamins, and most of the growth factors together with no adverse effects, as long as all soluble ingredients had been dissolved prior to autoclaving.

For sporulation medium, a simple medium containing 0.4% potassium acetate gives very good results with cells taken from a late log phase culture from a YEP dextrose medium. A more sophisticated medium is as follows:

Potassium acetate	1.0%
Bacto yeast extract	0.1%
Glucose	0.05%
Bacto agar	2.0%

Various presporulation media have been recommended. The author has not been convinced of the advantage of such media. In some cases, however, using a YEP medium with as much as 10% glucose or else with 3% glycerol instead of glucose or even addition of raffinose appeared to be of some use.

A medium used for diagnosing respiratory mutants by replica plating is a standard YEP medium with 3% glycerol substituted for glucose. If growth of respiratory-deficient mutants is desirable in plating tests, this medium contains 0.1% glucose in addition to glycerol. On such a medium, respiratory-deficient cells form small colonies.

Various dye indicator media have been used for identifying respiratory-deficient mutants. Their general applicability should be taken with a grain of salt (for reference, see Nagai[6]), and representative samples of colonies should be checked on YEP glycerol medium to test whether the indicator dye works reliably. Resistance against various fungitoxic agents is usually checked on a medium based on YEP dextrose, to which is added, normally after autoclaving, a sterile solution of the toxic agent.

The ability of yeasts to ferment certain carbon sources can be tested in a medium on a YEP basis to which has been added the respective sugar in the place of glucose. The most reliable method for this is to use Durham fermentation tubes, which trap CO_2 formed and thus indicate fermentation even if it is very slow and takes several weeks. During fermentation, acid is formed, which decreases the pH, a condition monitored by an appropriate dye: 0.4% bromcresol purple in 50% ethanol, 10 ml per liter for sucrose fermentation.

Basically, the very same media can be used for solid fermentation-indicator media in plates for replica plating. Reliable results can be obtained only under certain conditions since change in color is due to acid formation, and some strains produce acid even on glycerol medium, simulating fermentation otherwise absent on this medium; the rate of ferementation has to be fairly high, so that the acids formed are not used again. Therefore, solid fermentation-indicator media should be used only if the results thus obtained have been shown to be the same as with the fermentation tube assay.

B. Culturing of Yeast and Genetic Analysis

Yeast cells for all purposes are best grown on liquid YEP glucose medium under agitation. All strains used in genetic analysis of *Sacch. cerevisiae* grow very well even in the absence of a sufficient oxygen supply provided that the carbon source is glucose, sucrose, or another easily accessible sugar.

Some authors prefer to grow yeast for mutation experiments on solid media because, in haploid strains, cells seem to be better separated after growth on solid media than in liquid media.

Sporulation of diploid yeast cells is achieved both in well-aerated liquid and on the surface of solid sporulation media. Sporulation becomes visible after 12–14 hr, and well-developed asci obtain after about 3 days. The optimal temperature for yeast growth is between 25 and 35°C; for sporulation, the lower temperature is preferable.

A well-sporulated culture is used for genetic analysis. There are two ways of obtaining isolated haploid spores from an ascus: tetrad analysis and random spore isolation. Random spore isolation is very easy in *Schiz. pombe,* where asci disintegrate spontaneously and easily. Nonsporulated cells can be selectively killed by treatment with 50% ethanol.[5,7] *Sacch. cerevisiae* does not allow for this procedure. The ascus wall has to be disrupted artificially. This can be achieved by treating sporulated cultures with commercially available enzyme preparations. In Europe, the most popular source is snail enzyme (suc d'*Hélix pomatia* from Industrie Biologique Française S.A., Génnévilliers, France), and in the United States it is Glusulase (from Endo Laboratories, Inc., Garden City, N.Y.) These commercial preparations are diluted 1:5 or 1:10, and incubations are run for various lengths of time depending on strain, preparation and age of the product, and intended type of analysis.[8]

If random spore analysis is preferred, digestion is carried on until the ascus wall is completely dissolved. The lipophilic spores tend to cluster together and can be separated by ultrasonication or else can be absorbed into paraffin oil. It is hard to destroy selectively all nonsporulated cells. The paraffin method of Emeis and Gutz[9] allows for considerable enrichment of spores in poorly sporulated cultures. Well-digested cell populations in an aqueous suspension are mixed with paraffin oil. The paraffin–water emulsion is then separated by centrifugation, and paraffin phase is extracted several times with water to remove vegetative cells. Finally, spores in paraffin are streaked out on a YEP glucose media plate. All clones arising there should be checked for haploidy, and one has to be aware of the fact that at least some diploid, vegetative cells can be carried along. The novice usually considers this method to be generally superior to tetrad analysis as far as the amount of work is concerned; the expert prefers tetrad analysis, as long as sporulation is good.

Tetrad analysis on only partially digested cultures requires little time and has the advantage of reliability. Digestion has to be reduced to a limit which allows a relatively easy bursting of asci with a micromanipulator needle on an agar slab (3% agar in a 0.4% solution of potassium acetate).

It is debated which is the best micromanipulator to be used. Technically, however, there is no better one available than the de Fonbrune micromanipulator. A single joy stick allows one to move the needle in two directions in one plane, and a knob on top allows lifting and lowering the needle. The excursions of the needle can be adjusted as desired. With this manipulator, a patient person can dissect about 60 asci an hour. The disadvantage of this machine is that it is not firmly attached to the microscope, which should therefore have a fixed stage and a long working range objective. A total magnification of 200 X is usually enough. Various other mechanical manipulators are in use, but most of them have the disadvantage that in order to move the needle in all three dimensions more than one lever has to be moved. The advantage is that they can be attached to the microscope stage.

No matter which micromanipulator is used, the following procedures are followed: A needle is made which extends from a glass rod at a 90° or a 45° angle. The tip of this needle should have a diameter between five and ten times the diameter of an ascus. Sporulated and digested cultures are streaked on an agar slab on a wide glass slide. This is placed, agar down, on a frame so that the needle can be moved between the agar and the stage. Asci can be picked up by adhesive forces by the needle. The stage is then moved away from the streak, and the ascus is deposited about 5 mm away. Slight vibrations of the needle will cause the ascus to disintigrate. Individual spores are picked up again and deposited in a row (3 mm apart). After a slab has been completed, it is placed onto YEP glucose medium and incubated. In order to make sure that no random aggregates of four spores have been collected as an ascus, it is germane to have segregating in a cross markers other than the one whose segregation is to be followed. With some practice, tetrad analysis is faster than random spore isolation and, on top of that, much more informative.

C. Preparation of Synchronous Cultures

For some purposes, it is important to have synchronized cultures. Various methods have been advertised. The problem is always that some strains obey well and others do not, so several methods should be tried. Williamson and Scopes[10] described a method which has been quite successful in other hands.[11] This method is based on alternate feeding and starving of stationary-phase cells. A faster method has been described by Hartwell[12] based on periodic density fluctuations during the cell cycle. Cells of a log phase culture are mixed into a solution used to establish a gradient of Renografin and then centrifuged. Samples from the highest or lowest density are then used to inoculate a fresh medium in which the cells selected in this way will show synchronous growth (see also Sebastian et al.[13])

D. Maintenance of Stock Cultures

A major concern of geneticists is the maintenance of stock cultures. Yeast strains can be kept in a refrigerator in a YEP glucose medium for up to 1 year. It is important to store cultures in a medium which has not been exhausted. A good procedure is to put a small inoculum into fresh medium and store this in the refrigerator. Yeast grows very slowly in the cold, but viability is maintained for a long time. Permanent stock cultures can be obtained by absorbing cells to dried silica gel in the cold. The silica gel is dried at 150°C for 3 hr in small screw-cap vials, cooled down over P_2O_5, capped, and cooled in an ice bath before cells are added. Suspending cells in fat-free dry milk powder dissolved in a little water is suggested. However, good results can be obtained without this. After the cells have been absorbed, the solid cell–silica gel clump is broken, and the vial is closed, sealed airtight with paraffin, and stored in the refrigerator. Cells can be revived from such a silica gel culture by placing them in YEP glucose medium (and waiting for up to 2 weeks).

A most important point is to check a strain for the presence of the original markers, because many nutritional markers might reduce the growth rate a little, with the effect that upon continuous subculturing revertants to a prototrophic state will accumulate. In heterozygous diploids, markers might be lost due to mitotic recombination. For these reasons, all strains should be carefully checked before use. Haploid strains might accumulate diploid cells because of mutation of the mating type in one cell or as the result of other mechanisms. The resulting diploid will grow faster, and after several transfers the original haploid strain will have become diploid and no longer suitable for forward mutation to recessive nutritional requirements. A test on sporulation medium or abservation of increased cell size are easy criteria for detecting such changes.

III. TREATMENT CONDITIONS

The problems and techniques discussed in this section apply to all types of genetic alterations to be induced, from simple point mutations to cytoplasmic mutagenesis as well as to mitotic recombination.

If selective techniques can be applied, there are two basically different ways of treatment: incubation of cell suspension in a usually buffered solution of the agent to be tested, and treatment of cells after they have been spread onto solid medium in petri dishes.

A. Treatment in Suspension

Treatment in suspension has the advantage that cells are exposed to a well-defined concentration of a given agent for a defined length of time. Moreover, using buffered solution of an agent allows one to control the pH during the period of treatment.

An important point to be emphasized is the dosage problem. Theoretically, it might appear to be irrelevant whether the dose of a given agent is varied by changing the length of the time of treatment at a constant concentration or by changing the concentration at a constant treatment time. There are two factors that may influence the effect of these two types of dosage variation.

First, many mutagens are not very stable and decompose considerably even during a relatively short period of time so that the actual concentration used decreases during the treatment period. Therefore, dose–response curves have to be established using a constant period of treatment with the concentration being variable. The standard period of treatment has to be short relative to the half-life of the agent under the conditions used. This problem has been dealt with by Heslot.[14] Above this, there is another effect of time on the observed results. It is known that yeast cells can undergo recovery from lethal damage when held in non-growth-supporting suspensions[15]; this was studied thoroughly for genetic alterations by Parry and Cox[16] for ultraviolet light and by Zimmermann[17] for chemicals. This might be called the influence of the biological time factor.

The biological and the chemical time factors need not be very severe as long as one is interested in the genetic activity *v.s.* nonactivity at a qualitative level. For quantitative considerations, however, these factors have to be taken into account.

Very critical is the choice of *p*H. Most relevant is the *p*H range around neutrality. Some genetically active agents, such as nitrious acid, show a sharp response to *p*H. Other mutagens, however, show a clear *p*H dependence as well. [18]

The genetic effect of most chemicals shows a regular response to changes in temperature. This is reflected in an increase in activity with temperature. However, in some cases this normal response does not hold over a wide range; with increasing temperature during treatment, the genetic effect can reach a plateau or even decrease, as shown by Schwaier *et al.* [19] In the experiments described by these authors, N, N'-dimethyloxamide was definitely mutagenic at 15°C but virtually nonmutagenic 30°C. This compound is very labile, and it has to be kept in mind that some labile compounds decompose so fast that they might not reach the genetic compartment.

The sensitivity of cells to genetically active agents is usually quite variable during the cell cycle. The best method to solve this problem is to work with synchronized cultures. In many cases, this variation in sensitivity has only quantitative effects (see Holliday [20]), but qualitative effects may also be observed; i.e., a genetic effect may be exerted by a given agent only when it is administered during a specific stage of the cell cycle. Such a situation has been described by Esposito and Holliday [21] for the induction of mitotic recombination with 5-fluorodeoxyuridine. This chemical acts only when administered during the DNA-synthetic period of the cell cycle.

The appropriate choice of the cell-cycle stage is very important for those chemicals which might be mutagenic, not because of their reactivity with DNA, but because of their incorporation as base analogues or, as in the case of 5-fluorodeoxyuridine, their interference with specific reactions involved in DNA replication.

B. Spot Test on Plates

The spot-test procedure, where the agent is administered to cells on a plate, is very fast and simple. Fink and Loewenstein [22] plated auxotrophic cells onto a YEPD complex medium, administered the mutagen in the middle of the plate, allowed growth for 1 day, and finally replica plated onto a synthetic medium slective for reverse mutants. This procedure is very useful because it allows for enough growth to ensure the expression of the newly induced genotype. Reverted cells will continue to grow on the selective medium. This spot-test procedure can be used on cells plated directly onto selective media.[23] In the case of gene conversion, such a spot-test procedure has also been very efficient with cells plated on the selective medium immediately.[24] Because the spot-test procedure is very fast, the time-limiting factor is actually the preparation of solutions of the chemicals to be tested. For a more qualitative purpose, it is even enough to place a few crystals onto the plates previously seeded with the indicator cells. An example is shown for β-propiolactone, a liquid. It is important to have a complete-medium plate seeded with about 1000 cells to determine the range of killing created by the diffusion

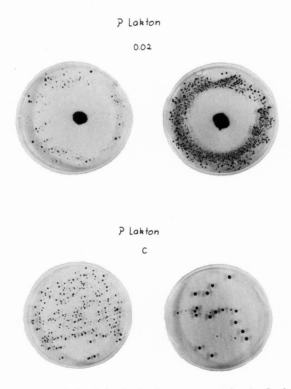

FIGURE 1. Spot-test assay: induction of mitotic gene conversion in *Saccharomyces cerevisiae* D4 (heteroallelic *trp5–12* and *trp5–27*). Cells of strain D4 were spread onto synthetic medium without tryptophan at a density of 10^6 cells per plate and onto synthetic complete medium at a density of 500 cells per plate. β-Propiolactone was added to the center of the plates (0.02 ml per plate). Killing can be observed on the complete-medium plate, whereas induction of mitotic gene conversion can be observed on the medium lacking tryptophan. Plates to the left, complete medium; plates to the right, medium without tryptophan. Top row, β-propiolactone; bottom row, control plates, with no additions.

of the chemical. Under appropriate conditions, the spot-test procedure yields very impressive results (see Fig. 1). However, a negative result in the spot-test assay does not necessarily indicate genetic inactivity. Trenimon, a very potent mutagen at least for yeast (see Fig. 2), is barely active in the spot-test assay. A quantitative conclusion is not to be derived from this procedure. It can serve a very useful prupose, however, for a fast survey of large numbers of chemicals or else for testing for the presence of a known genetically active agent. In the latter case, the spot-test assay can be readily quantitated. It is only the quantitative comparison of different chemicals which is difficult.

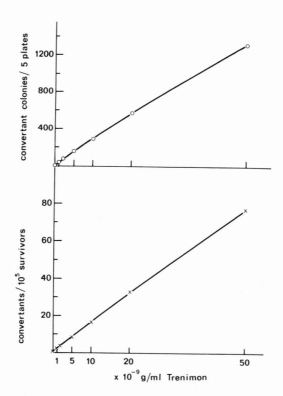

FIGURE 2. Example for a linear dose-response curve for the induction of mitotic gene conversion in *Saccharomyces cerevisiae* D4 (heteroallelic *trp5–12* and *trp5–27*). Stationary-phase cells were incubated with trenimon [2, 3, 5-tris(ethyleneimino)benzoquinone-4] in 0.1 M potassium phosphate buffer, *p*H 7.0, for 12 hr with shaking at 25°C. Upper graph, number of convertant colonies counted on five plates per dose. Lower graph, frequency of mitotic gene convertants per 10^5 survivors. Killing at highest dose 10.5%.

IV. SYSTEMS USED TO DETECT MUTAGENICITY OF CHEMICALS

Mutagenesis can be studied in *Sacch. cerevisiae* in haploid or in diploid cells. However, analysis is usually restricted to mutagenesis of the point-mutation type, whereas structural chromosomal mutations cannot be investigated easily due to the extremely small size of fungal chromosomes. Although it is a eukaryote, yeast cannot serve as a model organism for the induction of chromosomal aberrations using cytological methods. Of course, such alterations can be demonstrated by genetic crossing experiments, but this requires too much work to be useful for routine screening of chemicals. Induction of point mutations can be easily followed using selective or indicator techniques. The more popular methods used can be divided into forward mutations and reverse mutations, and these will be discussed here.

A. Forward Mutation

Forward mutations leading to a loss of function can be of two types: mutations leading to loss of an enzymic function required for biosynthesis, resulting in a growth factor requirement, and mutations leading to resistance to various toxic chemicals and substrate analogues.

1. Mutations Causing Nutritional Requirements

The first class of mutations is easy to obtain but not always easy to recognize. The best system to use is to look for red mutant colonies, which are due to a defect in adenine biosynthesis. The mutant colonies are easily recognized by their red pigmentation against a background of white colonies. In *Saccharomyces*, it is important to lower the adenine concentration to about half the usual amount in order to achieve good pigment formation. In *Schiz. pombe,* pigment is formed even in high-adenine media.[25] The genetic defects underlying this phenotype are located in the genes *ade1 and ade2* in *Sacch. cerevisiae* and in *ade6* and *ade7* in *Schiz. pombe.* Mutation frequencies based on scoring for the red phenotype are very high and can reach the percent level.[26] The advantage of this system over any other based on replica plating from a complete onto a minimal medium is that most mutants, at least at doses giving low killing, appear as red sectors in otherwise white colonies, and many sectors are very small. Using a dissecting microscope for screening plates will increase the efficiency so that the test system can be made very sensitive. Some authors plate only around 100–200 cells per plate, with the effect that the colony size is large and sectors are more conspicuous; other workers use many more, with the effect that the colony size is smaller and detection of small sectors is more difficult but fewer plates are required. This method is equivalent to the *ade3* system in *Neurospora* exploited by de Serres and Malling.[27] It has not been used in routine work for screening chemicals. However, it has provided means to study certain aspects of mutagenesis and has proven to be very helpful.[1]

Another use of the red adenine-requiring mutants is based on the fact that the blocks leading to the accumulation of the red pigment occur later in adenine biosynthesis. Consequently, any genetic block preceding the genes *ade1* and *ade2* and or *ade6* and *ade7*, respectively, for the two yeast species will prevent pigment formation. On top of that, mutation in other pathways may lead to a modification of pigment formation, at least in *Sacch. cerevisiae*. Red adenine-requiring cells are treated with mutagens and plated. Mutants show up as white colonies or colonies with white sectors against a background of red colonies. This system has been used by Roman[28] to study the genetics of adenine biosynthesis in *Sacch. cerevisiae,* and Leupold[7] established the same situation for *Schiz. pombe*. Since mutation of at least five gene loci can be detected starting with a red mutant, the optical yield of a mutation experiment is much better. Moderate doses of a good mutagen may yield white mutations at frequencies around 5%. Heslot[29] first used this system for mutagenesis experiments in *Schiz. pombe*, and in *Sacch. cerevisiae* this became quite popular, too. Marquardt *et al.*[30] used seven mutagens to test, as was previously done by Heslot, whether a given mutagen would show specificity for certain genes. In those experiments, white colonies or sectors were cloned and subjected to allelism tests *vs.* a set of tester strains representing those genes which functionally precede the red adenine genes. No clearcut case of mutagen specificity was observed in these experiments (for more references, see Mortimer and Manney.[1]

Theoretically, it is possible to use any other mutationally induced nutritional requirement as an indicator system for mutagenesis. Experienced workers have obtained dramatically high mutant yields.[31,32] However, the technique employed, replica plating from complete to minimal medium or streaking each colony before replica plating, is too tedious for routine work. Moreover, experiments using the mutational systems red adenine to white or wild-type to red adenine showed impressively that most mutants are found in sectored colonies. Most of these sectors are very small and would not be revealed by replica plating.

Snow[33] has described a method to enrich auxotrophic mutants in *Saccharomyces* and Megnet[34] a method for the same in *Schizosaccharomyces* using nystatin and 2-deoxyglucose, respectively. These methods have been useful in many hands in obtaining auxotrophic mutants, but, again, the amount of work involved does not suggest their use in routine mutagenesis screening.

In conclusion, there are a few practicable systems for auxotrophic mutations available which do not involve replica plating or other after-treatment manipulations and require only plating and visual screening of plates. Since the majority of auxotrophic mutations are recessive, work is restricted to haploid cells. There are published reports on mutagenesis in diploid yeast which started out with diploids heterozygous for red adenine mutations. The "mutation" frequencies observed are usually impressive but most of the time are due to mitotic crossing-over or gene conversion (see Zimmermann et al.[35]). Nevertheless, such a diploid system can be used to investigate diploid mutagenesis.

If the diploid cells treated are of the genotype ade2/+, they are still white due to the dominance of the wild-type allele. Mitotic crossing-over and gene conversion will give a stable homoallelic red diploid. The same phenotype can be created by mutation. However, the mutational ade2 homozygotes are not really homozygotes. Some of the mutant sectors are pinkish rather than dark red, and these cells quite often do not require adenine. In other cases, the sector may be perfectly red but unstable. Clones derived from such sectors throw off adenine-independent white papillae. These two properties are positive criteria to indicate that the red or pink sectors have arisen from mutation. If mutation occurs in an ade2-1/+ heterozygote and affects the ade2 wild-type allele, then this most likely does not affect the site which is affected in the ade2-1 allele but a different site. The result then will be a heteroallelic diploid in which the two alleles can intereact with each other at the genetic level to yield wild-type alleles by intragenic, conversional, or reciprocal recombination. This accounts for instability. It is known that certain ade2 alleles interact with each other at the protein level via allelic complementation[36] with the effect that a partially functional gene product is made. This accounts for pink or red clones not requiring adenine. This situation has been exploited by Zimmermann et al.[35] to study mutation in diploid cells. Mutations that delete part of or the entire ade2 gene will be stable and noncomplementing and cannot be recognized as mutations without further genetic analysis. Therefore, this system does not easily monitor all possible types of mutations as mutations but rather as unclassified genetic alterations.

The idea of using mutational systems based on loss of function rather than on restoration of function in reverse mutation experiments is that virtually all types of genetic alterations can lead to inactivation of a given gene. A high chance of detecting the full width of the spectrum of genetic alterations is provided by forward mutations from prototrophy to auxotrophy. The only mutations likely to be missed would be those where the genetic alteration leads to lethality, and these would be deletions covering more than a gene locus. This is the reason why mutagenesis is best studied in diploid cells, a condition which is complicated by the recessiveness of most auxotrophic mutations.

2.   *Mutations Causing Resistance to Toxic Agents*

Mutation to resistance can be studied on selective media containing a toxic substance. The advantage is that large numbers of cells can be plated out on a single plate where only mutated cells can grow. There are several agents known which are toxic and which can be tolerated by mutants. In some cases, the molecular mechanisms of resistance and the genetic basis are known: resistance to ions of copper,[37] cadmium,[38] and lithium,[39]; substrate analogues such as canavanine,[40] ethionine,[41, 42] p-fluorophenylalanine,[43] 5-fluorouracil,[44] and 2-deoxyglucose[34]; and finally to certain antibotics such as cycloheximide (actidione)[45] and nystatin.[46]

The well-understood cases of genetically determined resistance are those in which substrate analogues are involved which usually interfere with regulation of gene expression or enzyme activity. Such a resistance can be due to two basic mechanisms: either uptake of the analogue is impeded by mutation, or the regulatory properties of the mutants have been changed in a way to render them resistant to the action of the analogue. The lack of certain permeases involved in uptake of low molecular weight compounds which can also be synthesized by the cell usually leads to resistance to analogues. Any viable mutation preventing synthesis of a functional permease will then lead to resistance. Consequently, such a mutation system will monitor a broad spectrum of genetic alterations and appears, theoretically at least, appropriate for routine mutagenicity testing. Resistance based on alterations in regulatory mechanisms such as loss of feedback inhibition of a certain enzyme is restricted to mutations with mild effects such as missense mutations and will not be created by nonsense, frameshift, or deletion mutations. Therefore, any mutation system based on resistance should be understood in respect to the genetic and biochemical mechanisms; otherwise, one might use a narrow-spectrum mutation system without being aware of it.

According to Siegal and Sisler,[47] cycloheximide resistance is due to an alteration in ribosomal protein. A mutation in the respective structural gene(s) of the nonsense or frameshift type most certainly will lead to the formation of a completely nonfunctional protein which will be lethal. Therefore, only missense or very terminal nonsense or frameshift mutations can be expected to lead to the formation of a functional protein which causes resistance against cycloheximide. ICR-170 is considered to be a frameshift mutagen, and consequently it is not able to induce the required type of mutation.[48]

Brusick[48] demonstrated another important principle which interferes with resistance mutation. Canavanine resistance is due to an inactivation of an amino acid permease gene.[40] This mutation could be induced by ICR-170 as well as by the standard mutagen 1-methyl-3-nitro-1-nitrosoguanidine. However, mutants could be obtained only when stationary-phase cells were used; mutagenized cells taken from a culture during the logarithmic phase of growth did not yield resistant colonies when they were plated, immediately after treatment, on canavanine medium. This observation was explained by Brusick on the basis that cells during the logarithmic phase of growth had too many copies of intact permease so that canavanine was taken up even by those cells with a mutated permease gene. Therefore, they could not express the newly induced genotype. Nitrosoguanidine and ICR-170 are known to induce mutations in dividing cells as assessed by other mutation systems. The step of mutation expression is particularly important in those mutation systems which are based on the induction of resistant mutants.

Another difficulty with resistance mutation is that on the selective media colonies keep coming up for a long time. Mutagenized cell populations give rise to slow-growing colonies even on normal growth media, whereas colonies arise faster in the control population. This means, however, that the control population has more time to develop and accumulate spontaneous mutants than the slower-growing treated population. This results in an uncertainty about actual mutation induction over the control. Careful standardization of all conditions still allows for satisfactory results.[49] Another possibility of bias is the slow degradation of some analogues. Even with all these limitations, well-understood mechanisms of resistance can provide a useful mutation system. A still hypothetical resistance mutation system that avoids the complications by mutation expression and the appearance of "late" colonies would have the following properties: Sensitivity to a toxic agent is based on the activity of an enzyme which is subject to repression, e.g., to catabolite repression. This enzyme acts on a substrate analogue in a way to produce a highly toxic product. Cells used for mutation experiments are grown on a medium with 8% glucose so that synthesis of that enzyme is fully repressed. After treatment, these cells are plated on a medium with glycerol as a carbon source so that the cells start to synthesize this "toxifying" enzyme only after transfer to this medium. Those cells which have undergone an inactivating mutation at the respective structural gene locus for that enzyme or a regulatory gene will not be able to produce an active protein, with the effect that they are not inactivated by the toxic substance added to the glycerol medium. Those cells which become derepressed for the synthesis of this enzyme will be killed or inhibited due to the toxification of the substance. Allyl alcohol might be a substance suited for this purpose, since alcohol dehydrogenase converts this alcohol into highly toxic acrolein. There are two types of alcohol dehydrogenases in yeast, a constitutive enzyme and another one whose synthesis is repressed by glucose with the effect that the overall alcohol dehydrogenase activity is lower in glucose-grown cells. Mutants resistant to allyl alcohol are indeed affected in alcohol dehydrogenase.[50] Genetic analysis of such resistant mutants showed a normal Mendelian segregation even for dominant resistance. However, the results of mutation experiments were not clear, maybe due to an insufficiently large enough difference in alcohol dehydrogenase activity between glycerol- and glucose-grown cells,[51] but it might be possible to work out appropriate conditions for this toxic compound or others which are toxified by other repressible enzymes.

In summary, resistance mutation systems can be very useful and accurate, but the genetic and biochemical basis has to be well understood for a correct interpretation of mutation experiments based on their use. Otherwise, one might not detect certain types of mutations, as demonstrated by the work of Brusick.[48]

B. Reverse Mutation

Reverse mutation experiments have been very popular for studying chemical mutagenesis. Again, as in the case of resistance mutation, one has to be aware of the possibility that only certain types of mutations are leading to the reversion of the original defect. This, as a matter of fact, imposes severe restrictions on the use of reverse mutation as a means for monitoring genetic activity. Technically, reverse mutation systems are simple, accurate, and powerful. The number of cells plated on a medium dish should not be too high, due to the Grigg effect.[52] Nonrevertant cells unable to grow on a medium selective for revertants always show some residual metabolic activity or sometimes residual growth. This will lead to an exhaustion of growth factors in the medium to an extent that prevents true revertants from growing up to form a visible colony. On the other hand, large cell numbers per plate autolysing after doses giving high killing will release so many breakdown products that the few surviving and nonrevertant cells will grow on those. This situation can be demonstrated easily: selection plates seeded with cell populations with mostly dead cells usually develop small colonies of nonrevertant cells.

Besides these technical aggravations, there are basic obstacles in the use of reverse mutation systems. A true reverse mutation at the original mutant site requires a specific base-pair substitution, insertion, or deletion which need not be inducible by all mutagens. A very drastic demonstration of this situation is provided by the allele *ade6–45* of *Sacch. cerevisiae*. This allele has been used to detect the mutagenicity of a large number of mutagens of the nitrosamide type. These are thought to act as alkylating agents via hydrolysis to diazoalkanes.[53,54] The defect caused by mutation at the mutant site in this allele cannot be truly reverted or compensated by suppressor mutation[55] or induced by alkylating agents of the alkylsulfate and alkylsulfonate type.[56,57] Actually, all genetic experiments directed toward demonstrating base specificity of certain mutagens[58] are based on the fact that mutants induced with a given mutagen cannot be reverted by all mutagens, sometimes including the originally inducing mutagen. Of course, this situation can be largely overcome by using a whole battery of reverse mutation systems showing well-defined responses.[58,59] On the other hand, this situation allows one to determine at the molecular level the types of mutations that are induced by a new mutagen.

In conclusion, reverse mutation systems can be used with some caution to reveal mutagenicity of unknown chemicals. However, there are basic restrictions due to the usually specific genetic alteration leading to restoration of prototrophy. Nevertheless, it is this situation which allows one to gain a first idea about the types of mutations inducible with a new mutagen. For the practical purpose of detecting, with some degree of certainty, all types of mutational alterations, one faces the tantalizing situation that even a large battery of test systems might not be sufficient to monitor all possible types of genetic alterations.

C. Mutagenesis in Meiotic Cells

Cells undergoing meiosis differ considerably from normally growing mitotic cells. There are chromosome pairing, intrachromosomal and interchromosomal recombination, and reduction of chromosome number from the diploid to the haploid condition. All these additional reactions, as compared to what happens in mitotic cells, might contribute to additional proneness to mutation. Yeast is particularly suited for investigating such a problem. Using appropriate strains, an entire cell population can be induced to undergo meiosis more or less synchronously at a better than 80% efficiency.

It has been shown for *Saccharomyces*[60–62] and for *Schizosaccharomyces*[63] that cells undergoing meiosis are particularly prone to certain types of mutational changes, frameshift mutations above all. The procedure followed is to divide a population of mitotic cells into one fraction which is analyzed for its content of spontaneous mutants while the other fraction is induced to undergo meiosis, after which the asci formed are digested and ascospores are separated from each other by ultrasonication and finally tested for mutations. It could be shown, indeed, that certain chemicals such as acridine dyes are definitely mutagenic only when administered to meiotic cells and do not act on mitotic cells. On top of that, some acridine dyes, active mutagens in meiosis, have a protective effect in mitotic cells; i.e., they reduce the spontaneous mutation frequency.[64] These observations demonstrate a very important point: phase specificity of certain mutagens. It is believed that the meiotic effect is due to a specific action during the process of intrachromosomal recombination. It might be significant to mention here that Fahrig[65] was able to demonstrate a genetic activity for some acridine dyes in inducing mitotic gene conversion.

In conclusion, it must be emphasized here that inclusion of a test using meiotic cells is required if one intends to rule out mutagenicity of an agent in yeast. However, more work has to be done to elucidate and substantiate meiotic mutagenesis as a situation where chemicals not genetically active in mitotic cells exert a definite genetic effect.

D. Cytoplasmic Inheritance

Eukaryotic cells carry genetic information not only in the nucleus but also in organelles, such as mitochondria in yeast. In addition to that, there are cytoplasmic genetic factors known which cannot be associated with cell organelles, such as the killer factor[66,67] and the ψ factor.[68,69]

The most celebrated type of cytoplasmic genetic alteration is the genetically stable and irreversible change from a respiratory-competent to a respiratory-deficient state (abbreviated RC and RD, respectively). Certain yeast species, but by far not all, can be converted into such RD forms. The fascination about this possibility derives from the fact that certain chemicals such as acridine dyes[70] can induce such genetically stable variants at an almost 100% yield. Recognition of such altered forms is, or at least seems to be, very easy by plating cells on certain indicator media which allow for a ready distinction between RC and RD cells (see Nagai[6] for a variety of media recipes). There is no doubt about the fact that genuine mutagens induce a genetically stable and cytoplasmically inheritable change from a RC to a RD condition, as shown by subsequent genetic analysis (for references, see Schwaier et al.[71]). Most reports in the literature are devoid of demonstrations concerning the type of genetic alterations induced; the best argument presented in favor of the cytoplasmic as opposed to the nuclear type of mutations can be derived from the more or less deliberate choice of diploid instead of haploid cells. This and the usually extremely high yield of mutants can be taken as suggestive evidence for the cytoplasmic nature of the observed phenotypic alterations as monitored on a given type of indicator medium. If genetic analysis is performed adequately, a very complex situation is revealed, as clearly demonstrated by Sherman. [72,73]

Various regimens inducing the RD phenotype have been proven not to be mutagenic in any other system in yeast: KCN, [74] incubation in certain dilute media, [75] and growth at elevated temperature. [76] This shows very clearly that a change from an RC to a stable RD phenotype can be induced not only by definite mutagens.

The ease with which RD mutants can be obtained in certain yeasts [5,77,79] indicates only a very weakly established cell organelle system rather than a true mutability of cytoplasmic genetic factors. This situation was clearly documented by De Deken, [79] who showed that in all yeast strains investigated respiration could be inhibited by the acridine dye euflavine but only some species could be converted into stable RD forms.

In conclusion, it must be emphasized that the induction of the RD or petite phenotype (petite colony phenotype due to slower growth) can be brought about by mutagens as well as by definitely nonmutagenic regimens and, moreover, that induction of this alteration is possible in only a few yeast species. Consequently, results obtained using this "mutation" system cannot be considered as of general relevance.

Genetic alterations other than the RD phenotype can be induced in mitochondria. Certain antibacterial antibiotics inhibit mitochondrial function with the effect that yeast cells are unable to grow on nonfermentable carbon sources in their presence. Mutants have been obtained which are resistant to this antibiotic impediment of mitochondrial function, and the cytoplasmic nature of some of the mutants has been clearly established. [80,81] However, both nuclear and cytoplasmic factors can mutate to provide resistance. No use has been made of this system to establish a genetically sound cytoplasmic mutagenesis test system.

If the drug resistance level caused by cytoplasmic mutation is higher than that caused by nuclear gene mutation, one can specifically select for cytoplasmic mutants. This situation seems to be realized in the case of erythromycin resistance. [81] Clearly, this situation appears to provide a genetically sound test system, but it has to be calibrated before it is used for routine screening.

Yeast provides the possibility of testing for cytoplasmic mutagenesis. This possibility should not be neglected, because it is conceivable that there might be chemicals that act on cytoplasmic genetic factors but only weakly or not at all on nuclear genes or *vice versa* (see Schwaier *et al.* [71]). However, a thorough understanding of the physiology and genetics of such alterations is indispensable for a proper evaluation of the phenomena observed.

V. EVALUATION OF MUTAGENESIS EXPERIMENTS AND COMPLICATIONS

It must be borne in mind that mutagenesis is not just one reaction but a series of reactions. The first step is uptake or penetration of the chemical. A negative effect of a chemical when tested for genetic activity can be due to two facts: the compound cannot enter the cell, or else it is entering the cell but is not active at all. In the first case, no conclusion can be drawn as to the genetic activity. However, if an agent is genetically inactive but kills the cells of the test organism, one can assume that the chemical to be tested has entered the cells. In this case, lack of genetic activity means really no genetic activity in this organism under the conditions used. This is a relatively trivial problem, but it is not to be neglected.

A. Mutation Fixation and Mutation Expression

Provided a chemical has penetrated into the cell, it can alter DNA directly by reacting with DNA constituents, indirectly by being incorporated as a DNA precursor analogue, or else still more indirectly by interfering with systems maintaining DNA integrity. Quite often, the product of the primary alteration of the DNA molecule is not yet the form in which the new genetic information is transmitted in subsequent cell generation. The primary reaction is followed by the processes of mutation fixation (for review, see Auerbach[82]). It has been observed in yeast that after-treatment conditions can drastically influence the frequency of mutation. Temperature is one of the factors, postincubation at 25°C yielding higher mutation frequencies than 30 or 35°C. Such a temperature-sensitive period can extend over a period of almost 20 hr.[83]

After mutation, there is a phase of mutation expression. Mutation expression is very relevant in systems based on the induction of resistance mutation. Resistance at the permease level can be expressed only after functional permeases have been diluted out of the membrane; this applies to canavanine or other types of analogue systems.

As a general rule, the problem of expression has to be considered in all systems based on the use of selection of mutated cells. Particularly critical are experiments which are aimed at including DNA precursor analogue mutagenesis, because this involves the longest mutational pathway of all agents.

The most complicated situation is provided by those agents which are genetically active not because they react directly with DNA or are incorporated into DNA in the form of precursors but because they interfere with those enzymatic systems which maintain DNA integrity. The experimental setup used to detect such agents is based on challenging such "repair" systems by treatment with a known mutagen. Cells can be exposed to the agent to be tested before, during, or after treatment (see Auerbach.[85]). In the case of caffeine, effects have been tested by plating cells of *Schizosaccharomyces* treated with 1-methyl-3-nitro-1-nitrosoguanidine or ultraviolet light onto media containing 0.2% caffeine. Caffeine reduces the number of mutants and mitotic, intergenic recombinants as well as survival.[84]

One grave problem in mutation research is mutagen specificity. It has been discussed above that reverse mutation systems cannot be used reliably because the alteration selected for, restoration of prototrophy in an auxotrophic mutant strain, can be brought about only by a limited number of genetic alterations.[57] This type of mutagen specificity is very obvious. It might be argued that other systems based on induction of resistance or induction of nutritional mutants will be free of such restrictions. This is not the case. Auerbach[85] has published a thorough discussion of the problem of mutagen specificity. The complications caused by the phenomenon of mutagen specifity are severe and in most cases due to processes involved in mutation expression rather than due to the actual process of mutagenesis in terms of chemical alterations of DNA.

Forward mutation systems based on the inactivation of genes involved in the synthesis of a low molecular weight compound such as histidine would be considered as quite revealing, because any type of mutational alteration can cause a histidine requirement. Consequently, a system based on induction of histidine auxotrophs should not show any mutagen specificity. However, Zetterberg[86,87] could demonstrate in the filamentous fungus *Ophiostoma* a dramatic example of mutagen specificity. Spores treated with ultraviolet light or N-methyl-N-nitrosourethane were plated on complete medium, and colonies appearing there were tested for their nutritional requirements. Spores treated with ultraviolet light yielded many auxotrophic mutants, none of which required histidine. Histidine-requiring mutants could be obtained readily after nitrosomethylurethane treatment. Ultraviolet light induced histidine-requiring mutants, but they could only be recovered if the treated spores were plated on minimal medium supplemented with histidine; complete medium somehow inhibited the growth of ultraviolet-induced histidine mutants. Such plating-medium effects have been studied by Clarke[88] and also by Hénaut and Luzzati.[89] In the latter case, the genetic alteration studied was gene *converversion* at the *ade3* locus in *Saccharomyces;* addition of histidine to the plating medium prevented the appearance not only of prototrophs for *ade3* but also of prototrophs arising by gene conversion from a pair of heteroalleles of the *ura2* gene.

The best safeguard against such secondary effects on the expression or fixation of induced genetic alteration is a calibration of the system used with a wide variety of mutagens, and again, as in the case of resistance mutation, some understanding of the genetic and physiological effects of the genes mutated might be quite helpful. In the case of *ade3* gene conversion, it was found that the inhibiting effect of histidine was due to the complex regulatory function this gene has in methylation processes. However, the functional role of many genes is well understood, and complications of the types mentioned above can be easily avoided by looking for genetic alterations that influence simple functions in a dispensable pathway.

The problem of mutagen specificity as far as plating-medium effects are concerned has to be borne in mind for a proper evaluation of the data obtained. Actually, it is the complications introduced by these phenomena which make a simple cookbook type of approach to routine screening for mutagenicity impossible.

B.   Dose-Response Curves

Dose-response curves should always be established in routine testing. Many very efficient mutagens induce very high frequencies of genetic alterations long before killing becomes significant. Other mutagens have a strong killing effect and show less genetic activity. Such agents have a very narrow biologically relevant dose range; i.e., a twofold increase in dose can lead from almost no killing to a virtually complete inactivation of the entire cell population. In such a situation, a careful investigation of the biologically relevant dose range is necessary to detect a possible genetic effect.

Ideally, dose-response curves follow simple relations; they are either linear (Fig. 2) or else follow a "two-hit kinetics" function. Such situations can be found at lower dose ranges, but at higher doses curves can plateau or even decline. Extensive investigation of dose-response relationships with chemically induced mutation has been reported by Heslot[14] and Schwaier[55] and by Kølmark[90] in *Neurospora*.

Studies of variations in temperature and pH on mutation induction are sometimes difficult, because the shape of dose-response curves can vary considerably (see Schwaier[55]). At any rate, if effects of pH, temperature, or other factors on the genetic activity of a certain agent are to be investigated, this cannot be done using just one standard dose; rather, an entire dose-response curve has to be established in order to arrive at reasonable conclusions. Westergaard[91] has used the "optimal dose," the dose which gives the highest mutation frequency, as a basis of comparison.

It is very easy to establish the genetic activity of an agent that leads to an absolute increase in the number of mutants per treated cell because the rate of increase of mutants over spontaneous is greater than the rate of inactivation per unit dose increment. Weak mutagens, however, are difficult to evaluate because all that is observed sometimes is an increase in mutants per surviving cell but not per treated cell. Such an observation leaves room for two interpretations: the observed relative increase is due to induction of new mutants, or else such an increase is simulated by a higher resistance of spontaneous mutants to the inactivating effects of the agent tested. Reconstruction experiments can solve this problem. More complicated is the situation in which the cell population is heterogeneous with respect to susceptibility to spontaneous mutation and this susceptibility is correlated with increased resistance. There are no good reconstruction experiments to test for such a possibility. Therefore, weak genetic effects even if they are statistically significant do not necessarily mean that they are genetically significant, i.e., due to mutation induction. Such a situation might become very serious in those cases where an increase in dose is impossible due to the insolubility of the chemical to be tested. It is always the negative or ambiguous result which raises the problem of the validity of the test used.

VI. THE PROBLEM OF METABOLIC ACTIVATION

Yeasts are unable to activate various chemicals which are active mutagens or carcinogens in mammals. Nitrosamines are potent mutagens in *Drosophilia*[92] but absolutely inactive in *Neurospora*,[93] yeast,[53] and bacteria.[94] The same situation applies to aromatic amines. Marquardt *et al.*[95] tested a large number of aromatic amines and their presumptive metabolites formed in mammals. The results clearly demonstrated that the presumptive ultimate metabolic derivatives are active in inducing genetic alterations. Metabolic reactions occurring in mammalian metabolism can be mimicked in *in vitro* reactions such as oxidations or hydroxylations of various amines, as demonstrated by Malling[96] in *Neurospora* and Mayer[49] in yeast. Reactions *in vivo* include the host-mediated assay, which has been shown to be applicable to yeast as a microbial indicator organism by Fahrig[97] and Marquardt and Siebert,[98] the latter authors using rat urine as a source of highly concentrated active metabolites of endoxan, a compound inactive in yeast in its original form. Liver homogenates proved to be useful in activating chemicals which are inactive *per se* in *Neurospora*.[99] These conditions are very important because they allow for the use of indicator organisms which are easily amenable to a perfect genetic analysis. In the hands of an experienced worker, such methods should yield reliable results.

VII. MITOTIC RECOMBINATION

Mutation of nuclear or cytoplasmic material is not the only genetic alteration to be observed after mutagen treatment of diploid cells. Mitotic recombination mediated by crossing-over and gene conversion will lead to two effects of genetic relevance: homozygosis and consequently expression of recessive markers and, in the case of intragenic recombination between two mutant sites within the same gene, new alleles. Two defective alleles with mutational alterations at different sites will yield by intragenic recombination, which is mostly of the nonreciprocal gene conversion type, two new alleles: the wild-type allele free from inactivating mutational lesions and the double-mutant allele. Information about these types of genetic alterations and the systems used to demonstrate them has been provided by Mortimer and Manney.[1] Although not very popular, mitotic recombination of both types has proven to be a useful tool in assessing genetic activity of chemicals (for review, see Zimmermann[24,100]).

A. Mitotic Crossing-Over

It has to be emphasized here that in order to study mitotic crossing-over certain precautions have to be taken. First, extended incubation of diploid yeast cells in buffer might induce sporulation and consequently lead to the expression of recessive markers. A control sample which shows an increase of "mitotic" recombination, especially in acetate buffers, indicates induction of meiosis. To avoid this, one has to shorten the time of treatment and use another buffer.

Mitotic crossing-over is usually detected by the appearance in a heterozygous strain of recessive markers, either a recessive resistance marker or *ade1* or *ade2* red adenine mutant alleles. The mere observation of recessive "homozygosis" is not sufficient to demonstrate induction of mitotic crossing-over. The same phenotype can be created by (very unlikely in *Saccharomyces*) haploidization, aneuploidy, chromosomal deletion, mutation of the wild-type allele, gene conversion, and finally (and most frequently) mitotic crossing-over. In order to establish firmly the induction of mitotic crossing-over, one has to show the generation of the two reciprocal products (for discussion, see Zimmermann *et al.*[35]). Appropriate systems to allow for a complete analysis have been presented.[1,35] It is always advantageous to set up experiments in a way to avoid an immediate exposure of cells to selective conditions, even though this is more convenient, because of problems of expression of the new, homozygous genotype.

Demonstration of the two reciprocal products is a necessary and sufficient criterion to demonstrate induction of reciprocal, mitotic crossing-over. A very elegant and simple system has been suggested by Cox.[101] A diploid is established, heteroallelic for *ade2*, one allele being a normal *ade2* mutant allele expressing the typical deep red color when present in the haploid condition or in a homoallelic state in the diploid condition. The other allele has to be a leaky mutant which forms only little pigment but shows allelic complementation with the other allele to the extent that the resulting heteroallelic diploid will be white and not adenine requiring. Mitotic crossing-over will produce the two reciprocal products, which can be distinguished on the basis of their differences in pigmentation.*

Using mitotic crossing-over as a test system has definite advantages. According to currently held views (see Holliday[102]), damage to the genetic material might create a condition which results in mitotic recombination, and thus something like mutagen specificity is not very likely to interfere with the system unless some very unfortunate choice creates complications, as demonstrated for intragenic mitotic recombination in *ade3* diploids.[89] On top of that, an appropriate choice of markers should allow one to study mutagenesis in diploid cells as well, using the very same strain.

B. Mitotic Gene Conversion

Mitotic intragenic recombination is mostly due to gene conversion, and it can be induced with a large variety of mutagens. A strain heteroallelic at loci *ade2* and *trp5*, called D4, has been used to demonstrate genetic

---

*Such a strain has been established and can be obtained from the author.

activity for numerous chemicals. [17,24,65,97,98,103–105] The advantage of this system is that it allows for selective techniques and thus is simple, accurate, and sensitive; sensitivity can be enhanced by liquid holding. [17] Caution must be taken in order to avoid induction of meiosis during very long treatments. In addition to strain D4, other heteroallelic combinations have been used. [103,106,107]

VIII. METHODS USED TO INCREASE SENSITIVITY TO GENETICALLY ACTIVE AGENTS

Some chemicals show a very low solubility in aqueous solutions, which might impose serious restrictions on testing. An artificial increase in sensitivity is desirable. The use of cells during the logarithmic phase of growth might be indicated, and the use of synchronized cultures has shown that cells during DNA replication are more sensitive than cells in other phases of the cell cycle. [11,102] Another way to affect sensitivity for induction of mitotic gene conversion is liquid holding of treated cells before plating them on selective media. This has resulted in a sharp increase in mitotic convertants for ultraviolet-induced conversion [16] but in a reduction if the inducing agents were chemicals. [17] However, induction of mitotic gene conversion in respiratory-deficient heteroallelic diploids could be enhanced by liquid holding after treatment. [17,95] It is conceivable on the basis of these experiments that regimen's increasing sensitivity to some genetically active agents might not have the same effects on alterations induced by all chemicals.

Mutants sensitive to ultraviolet light or X-rays (cf. Cox and Parry [108]) can show in many cases a cross-sensitivity to chemical mutagens. [109] This suggests the use of such mutants in order to increase general sensitivity to genetically active agents. Such a possibility was put to test by Lemontt, [110,111] who used various ultraviolet light–sensitive mutants to induce various types of genetic alterations. This author convincingly demonstrated that only certain types of alterations could be induced more efficiently in sensitive mutants, whereas the inducibility of other types of mutations decreased sharply. The use of such sensitive mutant strains always implies the possibility that one might not detect all possible types of alterations, and consequently one might not be able to draw reliable conclusions.

In conclusion, various regimens and genetic conditions are known which make a given strain more susceptible to the induction of genetic alterations, and this has been used successfully by various authors. However, enough evidence points to the fact that such regimens and genetic conditions may not increase but rather decrease the inducibility of some types of genetic alterations. A positive effect gives a clear-cut answer, but negative results are ambiguous.

IX. GENERAL EVALUATION

Both yeasts, *Saccharomyces cerevisiae* and *Schizosaccharomyces pombe*, are simple unicellular organisms with genetic material arranged on chromosomes in a nucleus, in mitochondria, and in other as yet unidentified cytoplasmic structures. Genetic factors localized in the nucleus show a typical Mendelian segregation. Strains can be cultivated in the diploid as well as in the haploid phase. Any diploid cell can be sporulated under well-defined conditions, and any haploid cell can function as a gamete when brought into contact with another haploid cell of opposite mating type. This situation makes these yeasts ideal model organisms for the study of genetic alterations in eukaryotic, germline, and somatic cells. The only disadvantage is that the chromosomes are too small to be amenable to direct cytological observation.

Various mutation systems are available to test for mutation in haploid as well as in diploid cells in a way to allow for the detection of a very wide spectrum of mutational events. Furthermore, cytoplasmic mutation systems are available to detect genetic activity of those chemicals which might specifically act on cytoplasmic but not on nuclear genetic material.

Yeast cells can be induced to undergo meiosis under well-controllable conditions. This allows direct detection of genetic effects exerted specifically during the process of meiosis.

The stable diploid condition of *Sacch. cerevisiae* allows study of induction of mitotic recombination as mediated by crossing-over and gene conversion. Yeast cells cannot perform the same type of metabolic activation of chemicals that leads to mutagenicity and carcinogenicity in higher organisms of many compounds which are inactive in fungi. However, various types of metabolic activations can be mimicked by *in vitro* systems based on simple chemical reaction mixtures or else by use of cell-free extracts. Yeast has been used as a genetic indicator organism in the host-mediated assay or for treatment with urine from mammals injected with a chemical that requires activation by mammalian metabolism.

The wide variety of genetic alterations that can be induced and detected in yeasts makes these organisms very suitable for studying genetic effects of chemicals. On the other hand, proper evaluation of such a wide spectrum of genetic alterations requires a thorough knowledge of the genetics and physiology of these organisms. The major goal of this chapter has been to point out sources of complication and errors which might lead to erroneous conclusions.

X. ACKNOWLEDGMENTS

I wish to thank Dr. N. R. Eaton for his help in writing the manuscript. Discussions with Drs. D. J. Brusick and V. W. Mayer have led to a deeper penetration into the problems and possibilities encountered in the use of yeasts as genetic indicator organisms for chemical mutagenesis.

XI. REFERENCES

1. R. K. Mortimer and T. R. Manney, *in* "Chemical Mutagens" (A. Hollaender, ed.), Vol. 1, PP. 289–310, Plenum Press, New York (1971).
2. K. Esser, *Mol. Gen, Genet. 110,* 86–100 (1971).
3. H. Gutz, *J. Bacteriol. 92,* 1567–1568 (1966).
4. M. Flores da Cunha, *Genet. Res. 16,* 127–144 (1970).
5. K. Wolf, M.. Sebald-Althaus, R. J. Schweyen, and F. Kaudewitz, *Mol. Gen. Genet. 110,* 101–109 (1971).
6. S. Nagai, *J. Bacteriol. 86,* 299–302 (1963).
7. U. Leupold, *Arch. Julius Klaus-Stift. Vererb. Forsch. 30,* 506–516 (1955).
8. J. R. Johnston and R. K. Mortimer, *J. Bacteriol. 78,* 292 (1959).
9. C. C. Emeis and H. Gutz, *Z. Naturforsch. 13b,* 647–650 (1958).
10. D. H. Williamson and A. W. Scopes, *Exptl. Cell Res. 20,* 338–349 (1962).
11. R. E. Esposito, *Genetics 59,* 191–210 (1968).
12. L. H. Hartwell, *J. Bacteriol. 104,* 1280–1285 (1970).
13. J. Sebastian, B. L. A. Carter, and H. O. Halvorson, *J. Bacteriol. 108,* 1045–1050 (1971).
14. H. Heslot, *Abhandl. Deutsch. Akad. Wiss. (Berlin) Klins. Med.* 191–228 (1962).
15. M. H. Patrick, R. H. Haynes, and R. B. Uretz, *Radiation Res. 21,* 144–164 (1964).
16. J. M. Parry and B. S. Cox, *Mutation Res. 5,* 373–384 (1968).
17. F. K. Zimmermann, *Mol. Gen. Genet. 103,* 11–20 (1968).
18. F. K. Zimmermann, R. Schwaier, and U. von Laer, *Z. Vererbungsl. 97,* 68–71 (1965).
19. R. Schwaier, F. K. Zimmermann, and U. von Laer, *Z. Vererbungsl. 97,* 72–74 (1965).
20. R. Holliday, *Genet. Res. 6,* 104–120 (1965).
21. R. E. Esposito and R. Holliday, *Genetics 50,* 1009–1017 (1964).
22. G. R. Fink and R. Lowenstein, *J. Bacteriol. 100,* 1126–1127 (1969).
23. D. Pittman and D. Brusick, *Mol. Gen. Genet. 111,* 352–356 (1971).
24. F. K. Zimmermann, *Mutation Res. 11,* 327–337 (1971).
25. A. Nasim and C. H. Clarke, *Mutation Res. 2,* 395–402 (1965).
26. N. Nashed and G. Jabbur, *Z. Vererbungsl. 98,* 106–110 (1966).
27. F. J. de Serres and H. V. Malling, *in* "Chemical Mutagens" (A. Hollaender, ed.), Vol. 2, pp. 311–342, Plenum Press, New York (1971).
28. H. Roman, *Compt. Rend. Lab. Carlsberg Ser. Physiol. 26,* 299–315 (1955).

29. H. Heslot, *Abhandl. Deut. Akad. Wiss. (Berlin) Klin. Med.* 98–105 (1960).
30. H. Marquardt, U. von Laer, and F. K. Zimmermann, *Z. Vererbungsl.* 98, 1–9 (1966).
31. F. Lingens and O. Oltmanns, *Z. Naturforsch.* 19b, 1058–1065 (1964).
32. F. Lingens and O. Oltmanns, *Z. Naturforsch.* 21b, 660–663 (1966).
33. R. Snow, *Nature (Lond.)* 211, 206–207 (1966).
34. R. Megnet, *Mutation Res.* 2, 318–331 (1965).
35. F. K. Zimmermann, R. Schwaier, and U. von Laer, *Z. Vererbungsl.* 98, 230–246 (1966).
36. R. A. Woods and E. A. Bevan, *Heredity 21,* 121–130 (1967).
37. D. C. Hawthorne and R. K. Mortimer, *Genetics 45,* 1085–1110 (1960).
38. J. E. Middlekauf, S. Hino, S. P. Yang, G. Lindegren, and C. C. Lindegren, *J. Bacteriol.* 72, 796–801 (1956).
39. W. Laskowski, *Genetics 41,* 98–106 (1956).
40. M. Grenson, M. Mousset, J. M. Wiame, and J. Bechet, *Biochim. Biophys. Acta 127,* 325–338 (1966).
41. H. Cherest and H. de Robichon-Szulmajster, *Genetics 54,* 981–991 (1966).
42. H. de Robichon-Szulmajster and H. Cherest, *Genetics 54,* 993–1006 (1966).
43. J. J. Gits and M. Grenson, *Biochim. Biophys. Acta 135,* 507–516 (1967).
44. F. Lacroute and P. P. Slonimski, *Compt. Rend. hebd. Seanc. Acad. Sci. (Paris) 235,* 2172–2174 (1964).
45. D. Wilkie and B. K. Lee, *Genet. Res. 6,* 130–138 (1965).
46. K. A. Ahmed and R. A. Woods, *Genet. Res. 9,* 179–193 (1967).
47. M. R. Siegal and H. D. Sisler, *Biochim. Biophys. Acta 103,* 558–567 (1965).
48. D. J. Brusick, *J. Bacteriol. 109,* 1134–1138 (1972).
49. V. W. Mayer, *Mol. Gen. Genet. 112,* 289–295 (1971).
50. U. Lutstorf and R. Megnet, *Arch. Biochem. Biophys, 126,* 933–944 (1968).
51. F. K. Zimmermann, Unpublished data.
52. G. W. Grigg, *Aust. J. Biol. Sci. 11,* 69–84 (1958).
53. H. Marquardt, F. K. Zimmermann, and R. Schwaier, *Z. Vererbungsl. 95,* 82–96 (1964).
54. R. Schwaier, F. K. Zimmermann, and R. Preussmann, *Z. Vererbungsl. 98,* 309–319 (1966).
55. R. Schwaier, *Z. Vererbungsl. 97,* 55–67 (1965).
56. F. K. Zimmermann, H. Marquardt, and R. Schwaier, *in* "Genetics Today," p. 59, Proc. Internat. Congr. Genet., Pergamon, London (1963).
57. H. Marquardt, R. Schwaier, and F. K. Zimmermann, *Mol. Gen. Genet. 99,* 1–4 (1967).
58. C. Yanofsky, *in* "Chemical Mutagens " (A. Hollaender, ed.), Vol. 1, pp. 283–287, Plenum Press, New York (1971).
59. B. N. Ames, *in* "Chemical Mutagens" (A. Hollaender, ed.), Vol. 1, pp. 267–282, Plenum Press, New York (1971).
60. G. E. Magni and R. C. von Borstel, *Genetics 47,* 1097–1108 (1962).
61. G. E. Magni, *Proc. Natl. Acad. Sci. 50,* 975–980 (1963).
62. G. E. Magni and P. Puglisi, *Cold Spring Harbor Symp. Quant. Biol. 31,* 699–704 (1966).
63. J. Friis, F. Flury, and U. Leupold, *Mutation Res. 11,* 373–390 (1971).
64. P. P. Puglisi, *Mol. Gen. Genet. 103,* 248–252 (1968).
65. R. Fahrig, *Mutation Res. 10,* 508–514 (1970).
66. J. M. Somers and E. A. Bevan, *Genet. Res. 13,* 71–83 (1969).
67. E. A. Bevan and J. M. Somers, *Genet. Res. 14,* 71–77 (1969).
68. B. S. Cox, *Heredity 20,* 505–521 (1965).

69. B. S. Cox, *Heredity 26,* 211–232 (1971).
70. B. Ephrussi, "Nucleo-cytoplamic Relations in Microorganisms," Claredon Press, Oxford (1953).
71. R. Schwaier, N. Nashed, and F. K. Zimmermann, *Mol. Gen. Genet 102,* 290–300 (1968).
72. F. Sherman, *Genetics 48,* 375–385 (1963).
73. F. Sherman, *Genetics 49,* 39–48 (1964).
74. T. J. B. Stier and J. G. B. Castor, *J. Gen. Physiol. 25,* 229–233 (1941).
75. S. Nagai, *Mutation Res. 8,* 557–564 (1969).
76. F. Sherman, *J. Cell. Comp. Physiol. 54,* 37–52 (1959).
77. C. J. E. A. Bulder, *Antonie v. Leeuwenhoek 30,* 1–9 (1964).
78. H. Heslot, A. Goffeau, and C. Louis, *J. Bacteriol. 104,* 473–481 (1970).
79. R. H. De Deken, *J. Gen. Microbiol. 44,* 157–165 (1966).
80. D. Wilkie, G. Saunders, and A. W. Linnane, *Genet. Res. 10,* 199–203 (1967).
81. D. Y. Thomas and D. Wilkie, *Genet. Res. 11,* 33–41 (1968) .
82. C. Auerbach, *Proc. Roy. Phys. Soc. 29* (1966).
83. F. K. Zimmermann, R. Schwaier, and U. von Laer, *Z. Vererbungsl. 98,* 152–166 (1966).
84. N. Loprieno and M. Schüpbach, *Mol. Gen. Genet. 110,* 348–354 (1971).
85. C. Auerbach, *Trans. Kansas Acad. Sci. 72,* 273–295 (1970).
86. G. Zetterberg, *Hereditas 47,* 295–303 (1961).
87. G. Zetterberg, *Hereditas 48,* 371–389 (1961).
88. C. H. Clarke, *Z. Vererbungsl. 93,* 435–440 (1962).
89. A. Hénaut and M. Luzzati, *Mol. Gen. Genet. 111,* 138–144 (1971).
90. G. Kolmark, *Hereditas 39,* 270–276 (1953).
91. M. Westergaard, *Experientia 13,* 224–234 (1957).
92. L. Pasternak, *Arzenimittel. -Forsch. 14,* 802–804 (1964).
93. H. Marquardt, R. Schwaier, and F. K. Zimmermann, *Naturwissenschaften 50,* 135–136 (1963).
94. E. Geisler, *Naturwissenschaften 49,* 380–381 (1962).
95. H. Marquardt, F. K. Zimmermann, H. Dannenberg, H. -G. Neumann, A. Bodenberger, and M. Metzler, *Z. Krebsforsch. 74,* 412–433 (1970).
96. H. V. Malling, *Mutation Res. 3,* 537–540 (1966).
97. R. Fahrig, *Mutation Res. 13,* 436–439 (1971).
98. H. Marquardt and D. Siebert, *Naturwissenschaften 58,* 681–682 (1971).
99. H. V. Malling, *Mutation Res. 13,* 425–429 (1971).
100. F. K. Zimmermann, *Biochem. Pharmocol. 20,* 985–955 (1971).
101. B. S. Cox, Personal communication.
102. R. Holliday, *Genet. Res. 6,* 104–120 (1965).
103. F. K. Zimmermann and R. Schwaier, *Mol. Gen. Genet. 100,* 63–76 (1967).
104. F. K. Zimmermann, *Z. Krebsforsch. 72,* 65–71 (1969).
105. D. Siebert, F. K. Zimmermann, and E. Lemperle, *Mutation Res. 10,* 533–543 (1970).
106. F. K. Zimmermann, *Mol. Gen. Genet. 101,* 171–184 (1968).
107. A. Putrament and H. Baranoswka, *Mol. Gen. Genet. 111,* 89–96 (1971).
108. B. S. Cox and J. M. Parry, *Mutation Res. 6,* 37–55 (1968).
109. F. K. Zimmermann, *Mol. Gen. Genet 102,* 247–256 (1968).
110. J. F. Lemontt, *Mutation Res. 13,* 311–317 (1971).
111. J. F. Lemontt, *Mutation Res. 13,* 319–326 (1971).
General Reference: A three-volume book, "The Yeasts" (A. H. Rose and J. S. Harrison, eds., Academic Press, London), contains many articles with information relevant to the problems encountered in using yeasts as a tool in chemical mutagenesis.

CHAPTER 32

Total Reproductive Capacity in Female Mice: Chemical Effects and Their Analysis*

W. M. Generoso and G. E. Cosgrove

Biology Division
Oak Ridge National Laboratory
Oak Ridge, Tennessee

I. INTRODUCTION

The necessity of studying the effects of certain chemicals on mammalian reproduction before they are introduced into the human environment is now fully recognized. This recognition stems from the fact that mammalian reproduction is a very complex biological function, in which many different processes come into play. Detrimental effects on one or more of these processes can lead to alterations in the normal reproductive pattern and represent a health hazard that may or may not be detectable by the present testing systems. This concern was expressed by the Food and Drug Administration Advisory Committee on Protocols for Safety Evaluation: Panel on Reproduction Report on Reproduction Studies in the Safety Evaluation of Food Additives and Pesticide Residues (Friedman *et al.,* 1970). Consequently, a three-generation plan, which was designed to study effects of food additives and pesticides on a number of reproductive end points, has been adopted by that advisory committee.

*Research sponsored jointly by the Food and Drug Administration, Washington, D.C., and the U.S. Atomic Energy Commission under contract with the Union Carbide Corporation.

The aspect of mammalian reproduction discussed here is the fertility effects observed during the entire reproductive life span of young adult female mice that have received a single dose of a test chemical before reproduction (a single-generation procedure). Reduction in the number of young born from females chemically treated, as such, could be due to genetic damage to the oocytes, oocyte killing, or interference by the chemical with normal oogenesis, ovulation, fertilization, or maternal physiology. At present there is a dearth of information on the detrimental effects of chemicals that are not hormonal in nature on female fertility (Cattanach, 1959; Röhrborn, 1966; Röhrborn and Berrang, 1967; Bollag, 1953; Krarup, 1970; Generoso, 1969; Generoso and Russell, 1969; Generoso *et al.*, 1971*a, b*).

One of the major objectives in the development of mammalian test systems for the wide-scale screening of chemicals is to be able to integrate different studies in a single batch of test animals. This is very important inasmuch as mammalian studies are relatively expensive. Thus, in addition to fertility effects, results of a study on delayed pathological and survival effects in chemically treated female mice in the total reproductive capacity experiments are also discussed here.

From the integrated reproductive and delayed pathological and survival studies, the following important points are illustrated: (1) there is a wide range in the fertility effects of aklylating chemicals; (2) reduction in the fertility of treated females may be due to dominant lethal mutations, oocyte killing, or some unknown nongerminal toxic effects; (3) the responses of reproducing females to the delayed survival effects of a given compound differ markedly from those of nonreproducing females; and (4) female mice are more sensitive to the detrimental fertility effects of a given compound than are males.

II. OOCYTE DEVELOPMENT AND RESPONSES TO RADIATION EFFECTS

Essential in the analysis and interpretation of fertility results is a knowledge of the nature of the mouse ovary. The total reproductive capacity procedure, as already stated, covers the study of fertility effects throughout the entire posttreatment reproductive life-span of females. At the time of treatment, all germ cells with the exception of those that will be ovulated within a few hours (Russell and Russell, 1955) are at the primary oocyte state—specifically, at diffused diplotene. The arrest of all germ cells at this meiotic state occurs shortly after birth. The majority of the oocytes of young adults are morphologically still the same small oocytes found arrested after birth; they are small, primordial oocytes with very few granulosa cells attached to their surfaces. A small proportion of the oocytes in the population are in various stages of follicular growth, characterized by increased size, increased numbers of granulosa cells, and formation of antra in the advanced stages of development. Classifications of the stages in follicular development have been given by Oakberg (1966), Pedersen and Peters (1968), and Peters (1969). It should be noted that the length of time required for the development of an oocyte from the most immature to the most mature form still is not known. Since a determinate number of oocytes are formed in females before birth, and the adult ovary lacks stem cells comparable to spermatogonia, as the animal ages the oocyte pool progressively decreases due to normal ovulation and spontaneous atresia. Hence, any agent that damages the oocytes will hasten the depletion of the pool and can lead to reduced fertility.

The first information on the response of mouse oocytes to the mutagenic and cytotoxic effects of an agent came from radiation studies. Radiation can decrease the fertility of irradiated female mice by way of induced dominant lethal mutations or oocyte killing. The oocytes of adult mice differ in their sensitivity to both the genetic and cell-killing effects of radiation, depending on the stage of development. Immature oocytes are most sensitive to the cell-killing action of X-rays (Brambell and Parkes, 1927; Murray, 1931; Oakberg, 1958; Schugt, 1928), but resistance of oocytes increases with the stage of development of follicles (Oakberg, 1958). Because the immature primordial oocytes constitute the main bulk of the oocyte population, radiation can have permanent detrimental effects on the reproductive capacity of female mice. For instance, (101 × C3H)F_1 females, when irradiated as young adults, produce throughout their reproductive life an average of only about four litters after acute exposure to 50 R of X-rays, and about 1.1 litters after 400 R, as compared to about 15 litters for unirradiated controls (Russell et al., 1959). In contrast to the cell-killing response, the most mature oocytes, i.e., oocytes in graafian follicles a few hours prior to ovulation, are most sensitive to the dominant lethal effects of X-rays (Russell and Russell, 1955). It should become apparent that studies done so far on the effects of chemicals on mouse oocytes have been strongly influenced by the earlier radiation work.

III. GENERAL PROCEDURE

For single acute treatment, two groups of young adult female mice (about 10–13 weeks old) each from two strains are given a single dose of the test chemical. Immediately after treatment each female of one group is caged permanently with a young untreated male. This group will undergo the total reproductive capacity test. Breeding pens are checked throughout the reproductive life of the females for the presence of newly born mice, which are discarded immediately after they are scored. Checks should be made only when young are expected; pens should be examined daily during weekdays beginning 18 days after either pairing or appearance of a litter. This procedure in the mouse maximizes the number

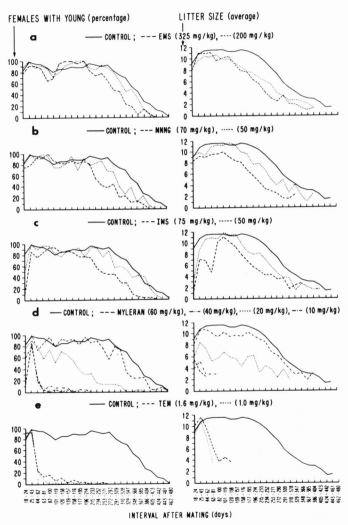

FIGURE 1. Fertility throughout reproductive life-span of female mice treated with the chemicals shown.

of young born to each mother. Effects of the chemicals on fertility can be analyzed using the following indices: (1) fertility of females throughout reproductive life—i.e., average litter sizes and percentages of females that give birth (also referred to as productive females) at time intervals that roughly approximate the normal succession of litters (Fig. 1), (2) average total number of young produced per female (Table 1), and (3) average number of litters produced per female (Table 1).

Figure 1 shows how data are summarized to show the reproductive pattern throughout reproductive life. The average litter sizes are plotted in the right-hand graph and percentages of productive females in the left-hand graph. In these graphs, the intervals were chosen on the basis of the length of the gestation period for the strain (about 18.5 days) and the length of the estrous cycle. Thus, most of the first litters in the controls are expected to appear within 18–24 days after pairing (first interval). In the later intervals, average litter sizes were plotted only when the number of productive females was three or more.

The second group of female mice is subjected to the dominant lethal mutation test. This test is necessary not only to analyze effects of chemicals on early litters but also because a small effect on fertility which might be detectable by an increase in dead implantation, for example, may not show up in the average size of first litters of females being tested for total reproductive performance. Females in the dominant lethal study are caged individually with untreated males on the day of treatment and are

TABLE 1. Reproductive Performance of Chemically Treated Female Mice

Treatment	Dose (mg/kg)	Number of females	Average interval to birth of first litter (days)	Mean number of litters produced per female	Mean number of young produced per female
Control		71	22.9	17.5	158
EMS	200	25	23.3	16.3	120[a]
	325	23	24.7	14.3[a]	106[a]
MNNG	50	24	22.3	14.2[a]	120[a]
	70	24	24.5	12.2[a]	89[a]
IMS	50	30	27.6	14.2[a]	114[a]
	75	59	34.2	11.3[a]	78[a]
Myleran	10	30	23.2	15.5[a]	131[a]
	20	31	23.6	8.0[a]	48[a]
	40	28	28.7	2.1[a]	9[a]
	60	25	35.4	1.3[a]	5[a]
TEM	1.0	30	22.8	2.2[a]	20[a]
	1.6	28	22.4	2.8[a]	26[a]

[a] $p < 0.01$ for difference from control.

checked for the presence of a vaginal plug (used as an indication of mating) every morning for 5 days. Mated females are separated from the males immediately after the observation of a vaginal plug and grouped together

TABLE 2. Histological Examination of Ovaries [a]

Chemical treatment	Treatment-to-fixation interval	Female identification code	Oocyte count per female[b]
Control	3 days	C-1	1222
		C-2	1775
		C-3	1831
	102 days	C-4	733
		C-5	908
	6 months	C-6	397
		C-7	547
	10 months	C-8	245
		C-9	257
EMS	3 days	E-1	1318
(325 mg/kg)		E-2	1588
	6 months	E-3	596
		E-4	743
	10 months	E-5	252
		E-6	196
MNNG	3 days	MN-1	1389
(70 mg/kg)		MN-2	1343
	6 months	MN-3	448
		MN-4	506
	10 months	MN-5	189
		MN-6	171
IMS	3 days	I-1	759
(75 mg/kg)		I-2	910
	102 days	I-3	115
		I-4	43
Myleran	3 days	My-1	1102
(40 mg/kg)		My-2	988
	7 days	My-3	720
		My-4	884
	14 days	My-5	208
		My-6	254
	120 days	My-7	0
		My-8	0
TEM	3 days	T-1	211
(1.6 mg/kg)		T-2	123
		T-3	177
	120 days	T-4	0
		T-5	0

[a] Females were of the (SEC × C57BL) F_1 hybrid strain.
[b] Sum of oocyte counts made on every fifth section, representing one-fifth of the oocytes.

according to the time of mating. Females are then killed 12–15 days after mating and scored for the number of corpora lutea, living embryos, and deciduomata. Presumed dominant lethal effects can be detected by using any of the following criteria: (1) increase in the frequency of dead implantation, (2) reduction in the average number of living embryos, (3) increase in preimplantation losses, and (4) reduction in the frequency of fertile matings. It should be emphasized that, singly or in combination, these criteria do not provide unequivocal evidence of true genetic effects. Since the females themselves received the treatment, it is very difficult to separate genetic from other effects.

When late fertility or sterilizing effects of the compound are observed, it is necessary to study the oocyte pool histologically. Ovaries from two of the females in the experimental and two in the control group are fixed as soon as effects are detected. Fertility effects can be detected quickly if reproductive data are summarized (as in Fig. 1) periodically. Serial sections (about 7 mμ thick) of the ovaries are prepared, and oocyte counts are made on every fifth section (Table 2). For all stages of follicular development, counts are restricted to oocytes with normal nuclei clearly showing in the section, independent of the general appearance of the follicle, since it has been found that destruction by a chemical of most of the granulosa cells in the surface of the oocyte does not necessarily result in oocyte killing (Generoso *et al.*, 1971*a*).

Throughout the entire reproductive life of the females, it is advisable also to obtain relevant delayed pathological and survival data. Females that die spontaneously during the test should be autopsied, and soon after the completion of the reproductive test each surviving female should be killed and studied for pathological abnormalities.

It should be noted that the total reproductive capacity procedure for female mice is simple and flexible. It is equally practicable for experiments requiring a single acute dose and for experiments requiring long-term exposures at chronic levels. With long-term exposures both male and female mice can receive the treatment, and when effects on fertility are observed the cause can be studied further. Also, the procedure can be combined easily with other tests, such as those for carcinogenicity and teratogenicity.

Whether acute or chronic exposure is involved, two important points must be kept in mind. First, the repeatability of the results, the sensitivity of the test, and the number of animals that should be used depend a great deal upon the strain of mice. Obviously, it is necessary to use females with good reproductive qualities. For a good strain of mice, such as (SEC × C57BL) F_1, 30 females per treatment are sufficient. And second, because the strain of mice is a very important factor in the testing of chemicals for mutagenicity in females (Generoso and Russell, 1969; Generoso *et al.*, 1971*a*), at least two unrelated strains must be used.

The hybrid strain (SEC × C57BL)F₁ exhibits suitable characteristics for the test. In this strain, most females mate immediately after parturition, and when the young are killed right after birth the litters are spaced approximately 19–20 days apart. The regularity of spacing is interrupted occasionally but usually returns. When data are pooled at suitable intervals, as in Fig. 1, the litters appear with remarkable regularity during the height of reproductive life. In the controls, the females skipped delivery only 1.2 times on the average during the first 12 intervals (the height of reproductive life). Another illustration of the desirability of the

TABLE 3.   Repeatability of Reproductive Performance of (SEC × C57BL)F₁ Females

Treatment	Dose (mg/kg)	Batch	Number of females	Average interval to birth of first litter (days)	Mean number of litters produced per female	Mean total number of young produced per female
Control	—	I	24	23.6	17.9 ± 0.3	157 ± 2.7
		II	29	22.7	17.3 ± 0.4	158 ± 4.1
		III	18	22.2	17.1 ± 0.5	161 ± 4.2
IMS	75	I	30	33.7	11.7 ± 0.5	82 ± 4.0
		II	29	34.8	10.9 ± 0.6	75 ± 5.2

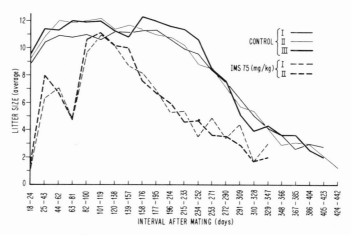

FIGURE 2.   Repeatability of fertility patterns throughout reproductive life-span of (SEC × C57BL) F₁ female mice.

(SEC × C57BL)F_1 strain can be seen in Table 3 and Fig. 2, showing data obtained from three batches of control mice and two batches of mice treated with 75 mg/kg of isopropyl methanesulfonate at different times. The three control batches involved 18, 24, and 29 females, while the two experimental groups involved 29 and 30 females. As can be seen, the three controls exhibited remarkably similar results, as did the two experimental groups.

IV. FERTILITY EFFECTS OF ALKYLATING CHEMICALS ON FEMALE MICE

Alkylating chemicals differ markedly in their ability to affect the reproductive performance of female mice. To demonstrate this, results with five alkylating chemicals are described here: triethylenemelamine (TEM), ethyl methanesulfonate (EMS), isopropyl methanesulfonate (IMS), N-methyl-N'-nitro-N-nitrosoguanidine (MNNG), and 1,4-di-(methanesulfonoxy)butane (myleran). Reproductive results for all these chemicals have been published elsewhere (Generoso et al., 1971b). Their effects on total reproductive performance of (SEC × C57BL)F_1 female mice are summarized in Table 1 and Fig. 1. It might be pointed out that in these studies all females were paired with normal males 1 day after treatment, and that the highest dose for each chemical is the maximum that could be tolerated by the animals.

At first glance, TEM would appear to have no effect on the early fertility of female mice, as two normal litters were produced by treated females. This may lead to the conclusion that TEM does not induce dominant lethal mutations in female mice. It should be noted, however, that none of the first litters shown in Fig. 1 resulted from matings that occurred within half a day after treatment, since in this particular study females were not mated for 24 hr. It has been observed that TEM induces low levels of dominant lethal mutations in the (SEC × C57BL)F_1 strain (Generoso et al., 1971a) and that most of these genetic effects are induced in oocytes ovulated within half a day after treatment (Generoso, unpublished data). Thus most of the dominant lethal effects were excluded, accounting for the observation that litters in the first interval were of normal size. Dominant lethal effects on three other strains have also been demonstrated (Cattanach, 1959; Generoso et al., 1971a). The small dominant lethal effect of TEM led to the conclusion that females should be caged with males immediately after treatment, instead of 24 hr later, and that whether or not a reduction in litter size is observed a dominant lethal study should be simultaneously performed (see Section III). The most obvious effect of TEM on fertility of female mice is the early, permanent cessation of reproduction, after an average of fewer than three litters per female. This rapid onset of sterility among TEM-treated females is due to destruction of oocytes in the early stages of follicular development (Table 2). A similar drastic effect of TEM on female fertility was observed by Cattanach (1959). TEM, therefore, can affect total reproductive performance of female mice by inducing dominant lethal mutations and oocyte killing.

Myleran can also produce both dominant lethal and cell-killing effects on mouse oocytes. Compared to TEM, myleran is much more effective in inducing dominant lethal mutations, and unlike TEM, with which dominant lethal effects were restricted to a short treatment-to-fertilization interval, myleran can induce dominant lethal mutations on oocytes ovulated 19.5 days or more after treatment (Generoso et al., 1971a). Because of its effectiveness in inducing dominant lethal mutations, myleran can cause litter-size reductions in the early litters, particularly in the first two, and temporary sterility among some females at higher doses. In addition to genetic damage, myleran is also effective in depleting the oocyte pool. As with TEM, high doses of myleran can cause early cessation of reproduction due to oocyte killing. Although no data are available on oocyte counts at lower doses of myleran, it is presumed that low fertility of treated females, which persisted throughout reproductive life, was due to some degree of oocyte killing commensurate with dose. A similar destruction of oocytes was observed by Bollag (1953) in myleran-treated rats, and it is interesting to point out that Bollag's study was prompted by his observation that a 42-year-old woman injected with myleran became amenorrheic 8 weeks after treatment. Although both TEM and myleran can effectively deplete the oocyte pool, there may be a difference in the mechanism by which oocyte killing is effected (Generoso et al., 1971b). TEM is similar to X-irradiation in that it induces a rapid destruction of oocytes that are in the earliest stages. After a high dose of TEM (1.6 mg/kg), all primordial oocytes (stage I in Oakberg's classification) are already gone within 72 hr. On the other hand, after a dose of myleran (40 mg/kg) that induces a fertility effect in terms of total number of litters produced comparable to that of 1.6 mg/kg of TEM, the degree of depletion of oocytes is much smaller 3 or 7 days after treatment than that observed 3 days after treatment with TEM. The degree of depletion of primordial oocytes 3 days after treatment with TEM is comparable to that observed 14 days after treatment with myleran. Thus, myleran seems to have a delayed oocyte-killing action.

Still a different pattern of dominant lethal and oocyte-killing effects is exhibited by the compound IMS (Generoso et al., 1971a, b). At a high dose (75 mg/kg), IMS-treated females showed a unique reproductive pattern characterized by an initially low fertility, followed by recovery and then a precipitous decline toward the latter part of reproductive life. The initially low fertility with this dose of IMS can be attributed to both dominant lethal and cell-killing effects on oocytes in large follicles. The cell-killing effect is more important in the reduction of fertility in the first interval, but in the second to fourth intervals dominant lethal effects are more pronounced. With the lower dose (50 mg/kg), the initial low fertility can be attributed mainly to dominant lethality (Generoso *et al.*, 1971a). The cell-killing effect of IMS on oocytes in the most advanced stages of follicular development is an interesting phenomenon, as the dose of 75 mg/kg does not destroy all the primordial oocytes. Thus, contrary to TEM, myleran, and X-rays, IMS appears to have a broad spectrum of oocyte-killing effects—i.e., both the very mature and the very young oocytes are affected. As already indicated, the reduction in fertility later in reproductive life is attributable to the cell-killing effects of IMS on primordial oocytes (Table 2). It should be emphasized that a large strain difference exists in the sensitivity of female mice to both dominant lethal and cell-killing effects of IMS (Generoso *et al.*, 1971a).

The effects of EMS and MNNG on the total reproductive performance of (SEC × C57BL)F_1 females demonstrate the complexity involved in analyzing chemical effects on fertility, as well as the real possibility that fertility effects may be an indirect manifestation of damage not directly involving the germ cells. The two chemicals are similar in that, at the highest doses, effects on this particular strain become pronounced only after several normal litters have been produced. These effects are similar to the late effects of the high dose of IMS. In contrast to IMS, however, the reductions in fertility induced by EMS and MNNG are not attributable to oocyte killing, as neither has any effect on the oocyte population (Table 2). Nor is there any evidence that these fertility reductions are due to dominant lethal mutations (Generoso, unpublished data). Thus it appears that the late effects of EMS and MNNG may in fact result from another type of damage, which presumably does not involve the oocytes directly.

EMS has no apparent effects on the early fertility of (SEC × C57BL) F_1 females, and a detailed study on this particular strain has indicated an absence of dominant lethal damage (Generoso and Russell, 1969). In another strain of female mice (T stock), however, EMS was found to be highly effective in inducing dominant lethal mutations in oocytes in the advanced stages of follicular development (Generoso, 1969; Generoso and Russell, 1969). This large strain difference obviously suggests that the total reproductive capacity test should be performed in at least two strains of female mice.

V. DELAYED PATHOLOGICAL AND SURVIVAL EFFECTS IN CHEMICALLY TREATED FEMALE MICE IN THE TOTAL REPRODUCTIVE CAPACITY EXPERIMENT

In addition to effects on reproduction, animals were also observed for late toxic effects. The first observation was a shortening of the life-span of treated females, which varied according to the chemical and dose used. Drastic reductions were observed among the MNNG- and IMS-treated females (by 30–42% and 19–23%, respectively), while a moderate reduction of the life-span was noted among females that received the high dose of EMS. As a result of this observation, early and late pathological effects were studied separately for a compound that had very little effect (EMS) and for one that had a marked effect (MNNG) on life-span. These effects were evaluated by necropsy of mice that either were killed periodically (sacrifice series) or died spontaneously (life-span series). MNNG or EMS at the dose levels used had no demonstrable histological effects in the early days and weeks after treatment. The results shown in Table 4 clearly confirm the marked life-span shortening with MNNG. The two compounds were also found to be potent inducers of lung tumors, but the incidence of tumors of other organs and lymphomas was not increased.

A very interesting observation is that among the MNNG-treated females those that were reproducing had a markedly shorter life-span than those that were not (Table 4, Fig. 3). (The nonreproducing females were caged with vasectomized males.) This observation holds true for

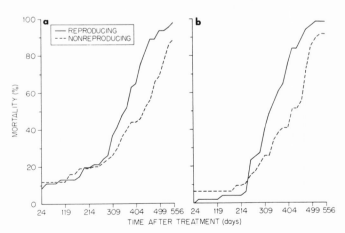

FIGURE 3.   Cumulative mortality of MNNG-treated female mice. (a) Strain (101 × C3H)F$_1$ (b) Strain (SEC × C57BL)F$_1$.

TABLE 4. Summary of Pathology Experiments

Experiment	Mouse strain	Number of mice	Mean survival time[a]	Acute tissue toxicity	Pathological findings (% incidence)		
					Lung tumor	Ovarian tumor	Leukemia lymphoma
Sacrifice series	(SEC × C57BL)F$_1$						
MNNG (70 mg/kg)		27	—	No	59	—	24
EMS (325 mg/kg)		21	—	No	67	—	0
Control		30	—	—	16	—	11
Life-span series, EAF[b]	(SEC × C57BL)F$_1$						
MNNG (70 mg/kg)		47	574	No	65	22	15
Control		33	798	—	46	30	60
Life-span, MGC[c]	(SEC × C57BL)F$_1$						
MNNG (70 mg/kg)		39	337	No	84	13	11
Reproducing	(101 × C3H)F$_1$	41	370	No	76	11	0
	(SEC × C57BL)F$_1$	34	443	No	87	16	10
Nonreproducing[d]	(101 × C3H)F$_1$	31	457	No	79	7	3

[a] Excluding deaths within 90 days of treatment.
[b] EAF, Experimental Animal Facility Colony, Biology Division, Oak Ridge National Laboratory. Females in this group were caged together.
[c] MGC, Mammalian Genetics Colony, Biology Division, Oak Ridge National Laboratory.
[d] Females were caged individually with vasectomized males.

two strains of females. In the ascending, midperiod portion of the mortality curves, the life-span of the reproducing females was estimated to average about 56 days shorter in the $(101 \times C3H)F_1$ strain and about 69 days shorter in the $(SEC \times C57BL)F_1$ strain than that of the non-reproducing females. The difference is even larger when the average survival times are compared. A preliminary perusal of the pathological findings in mice of the various groups does not reveal any reason for this difference. These results seem to imply that, for certain chemicals, reproducing females may respond differently to toxic effects than non-reproducing ones; and this may have important implications in the safety evaluation of chemicals.

VI. IMPORTANCE OF USING FEMALES IN FERTILITY STUDIES

We have attempted to show that studies on females, in addition to those on males, should be incorporated in the assessment of fertility effects of chemicals. Fertility effects in females may include a wide variety of possible health hazards such as mutagenic, cytotoxic, and other nongerminal undesirable effects that are not easily identifiable. Another reason for including females in mammalian fertility studies is clearly illustrated by our recent results with the compound hycanthone (Sterling-Winthrop, Rensselaer, N.Y.). (A full report will be published elsewhere.) This drug is currently being used in clinical trials to treat schistosomiasis, specifically in the control of *Schistosoma mansoni* and *Schistosoma haematobium*. We have found that hycanthone in the form of methanesulfonate, given as a single dose of 150 mg/kg, does not induce any detectable increase in dominant lethal mutations or any other fertility effects in male mice from two strains. (A few of the animals died within 2 hr after treatment.) Similarly, when the drug was given repeatedly, at 125 mg/kg per day for 5 successive days, no increase in dominant lethal mutations was detected. However, sterility was observed in matings that occurred on days $37\frac{1}{2}-41\frac{1}{2}$ after the last treatment. This sterile period corresponds to the period of differentiating spermatogonia and presumably resulted from the cytotoxic effect of the drug. The dominant lethal tests were performed on all stages in spermatogenesis. Since the dominant lethal procedure in male mice is a generally accepted test system in the evaluation of mutagenic hazards of chemicals, hycanthone could be declared not mutagenic in mice on the basis of these results. Our results on three strains of female mice, however, reveal that hycanthone has a marked effect on fertility. Studies on the effects of hycanthone on the total reproductive capacity of two strains of female mice are currently in progress.

Results so far show that a dose of 175 mg/kg (females seem to be slightly more resistant to the early toxic effects than males) induces a marked reduction in the size of the first litter in the two strains. Table 5 shows data for the (SEC × C57BL)F_1 strain. The litter-size reduction was analyzed in detail using a third strain, (C3H × C57BL)F_1, and three doses of the drug, 175, 150, and 100 mg/kg. In this study, treated females mated within 4.5 days after treatment (these matings correspond to the litters in the first interval) were killed between the twelfth and fifteenth days of pregnancy for uterine analysis. It was found that with all doses there were marked increases in the frequency of dead implantations. Obviously, the question that needs to be asked, which is very difficult to answer at the present time, is what causes the increase in dead implantations and in turn lowers the size of the first litters. One possible explanation is that the increase in dead implantations may be due to the early toxic effects of the drug, since at two higher doses some animals died within 2 hr after treatment. But even if no deaths occurred as with the 100 mg/kg dose, this explanation might hold true. Another possibility is that dominant lethal mutations are induced in oocytes in large follicles, since chemicals known to be mutagenic produce similar responses at certain doses. It must be emphasized that the distinction between genetic and toxic effects as the cause of the increase in dead implantations is very difficult to make since, unlike in male experiments, the females themselves received the treatment. However, regardless of whether this fertility effect is genetic in nature, one observation is beyond doubt: in the mouse strains studied, hycanthone exhibited a detrimental effect on fertility in females but not in males. The situation is reversed

TABLE 5. Effects of Hycanthone on the Early Litters of (SEC × C57BL)F_1 Females

Treatment	Treatment-to-birth interval (days)	Number of treated females	Number of productive females	Average litter size
Hycanthone	18–24	32	25	6.4[a]
(175 mg/kg)	25–43	32	29	10.9
	44–62	32	30	11.4
	63–81	32	26	11.8
Control	18–24	53	39	9.0
	25–43	53	50	10.6
	44–62	53	52	11.0
	63–81	53	48	11.3

[a] $P < 0.01$ for difference from control.

with EMS, MMS (methyl methanesulfonate), and PMS (*n*-propyl methanesulfonate), in that the males of certain strains show marked fertility effects due to dominant lethal mutations, while the females do not (Generoso and Russell, 1969; Generoso *et al.*, 1971*a*). Before hycanthone, all chemicals that had been shown to cause increases in dead implantations in females also produced dominant lethal effects in males, leading to the understanding that dominant lethal studies in males alone were sufficient and that nothing would be gained from similar studies in females. The hycanthone results, however, clearly demonstrate the need for the inclusion of female studies in the screening of mutagenic chemicals. Although its cause cannot be determined with certainty, the fertility reduction observed in treated females should be regarded as a possible health hazard and, if possible, analyzed accordingly.

VII. NEED FOR ASSAY SYSTEMS IN FEMALE MICE TO MEASURE INDUCED HERITABLE GENETIC DAMAGE

Direct proof for the chemical induction of heritable chromosome aberrations in mouse oocytes still is not available. Although certain chemicals are very effective in bringing about pre- and postimplantation death (dominant lethal–type response) in female mice, only indirect evidence is presently available for a genetic basis of damage to oocytes. This indirect evidence includes (1) the observation that preimplantation loss of embryos following prefertilization treatment of female mice with EMS is associated with an increase in the frequency of subnuclei among embryos in early cleavage stages (Generoso, 1969), (2) a later finding that IMS causes aborted meiosis among oocytes in the advanced stages of follicular development, leading to cell death (Generoso *et al.*, 1971*a*), (3) the recent finding that a significant increase in the frequency of chromosome aberrations is detectable among two-cell embryos from female mice treated with 2,3,5-triethyleneiminobenzoquinone-1,4(T) shortly before induced ovulation (Röhrborn *et al.*, 1971), and (4) the finding that cytogenetic damage caused by the antibiotics streptonigrin and phleomycin in oocytes has been detected through the presence of several types of abnormalities in treated oocytes analyzed cytogenetically at metaphases I and II (Jagiello, 1967, 1968).

These four findings by no means constitute unequivocal evidence of genetic hazards. Obviously, the most reliable evidence of mutagenic hazards from any agent in mice is the production of heritable mutations, which directly represent hazards. Thus it would certainly be highly desirable to determine the relationship of the presumed dominant lethal responses and the observed cytogenetic damage in females to heritable chromosome damage. Our inability to determine whether the effect of hycanthone in female mice is due to induced genetic damage clearly illustrates the necessity for further studies of the basic mechanisms involved in the chemical induction of chromosomal aberrations in female mice, as well as the need for a more reliable genetic assay system for the evaluation of mutagenic hazards of chemicals in mouse oocytes.

VIII. REFERENCES

Bollag, W. (1953), Der Einfluss von Myleran auf die Keimdrüsen von Ratten, *Experientia* *9*, 268.

Brambell, F. W. R., and Parkes, A. S. (1927), Changes in the ovary of the mouse following exposure to x-rays. Part III. Irradiation of the nonparous adult, *Proc. Roy. Soc. Ser. B 101*, 316–327.

Cattanach, B. M. (1959), The effect of triethylene-melamine on the fertility of female mice, *Internat. J. Radiation Biol. 3*, 288–292.

Friedman, L., Kunin, C. M., Nelson, N., Whittenberger, J. L., and Wilson, J. G. (1970), Panel on reproduction report on reproduction studies in the safety evaluation of food additives and pesticide residues, *Toxicol. Appl. Pharmacol. 16*, 264–296.

Generoso, W. M. (1969), Chemical induction of dominant lethals in female mice, *Genetics 61*, 461–470.

Generoso, W. M., and Russell, W. L. (1969), Strain and sex variations in the sensitivity of mice to dominant-lethal induction with ethyl methanesulfonate, *Mutation Res. 8*, 589–598.

Generoso, W. M., Huff, S. W., and Stout, S. K. (1971a), Chemically induced dominant-lethal mutations and cell killing in mouse oocytes in the advanced stages of follicular development, *Mutation Res. 11*, 411–420.

Generoso, W. M., Stout, S. K., and Huff, S. W. (1971b), Effects of alkylating chemicals on reproductive capacity of adult female mice, *Mutation Res. 13*, 171–184.

Jagiello, G. M. (1967), Streptonigrin: Effect on the first meiotic metaphase of the mouse egg, *Science 157*, 453–454.

Jagiello, G. M. (1968), Action of phleomycin on the meiosis of the mouse ovum, *Mutation Res. 6*, 289–295.

Krarup, T. (1970), Oocyte survival in the mouse ovary after treatment with 9, 10-dimethyl-1, 2-benzanthracene, *J. Endocrinol. 46*, 483–495.

Murray, J. M. (1931), A study of the histological structure of mouse ovaries following exposure to roentgen irradiation, *Am. J. Roentgenol. Radiation Therap. 25*, 1–45.

Oakberg, E. F. (1958), The effect of X-rays on the mouse ovary, *Proc. Internat. Congr. Genet. 10*, 207.

Oakberg, E. F. (1966), Effect of 25 R of X-rays at 10 days of age on oocyte numbers and fertility of female mice, in "Radiation and Ageing" (P. J. Lindop and G. A. Sacher, eds.), pp. 293–306, Taylor and Francis Ltd., London.

Pedersen, T., and Peters, H. (1968), Proposal for a classification of oocytes and follicles in the mouse ovary, *J. Reprod. Fert. 17*, 555–557.

Peters, H. (1969), The development of the mouse ovary from birth to maturity, *Acta Endocrinol. 62*, 98–116.

Röhrborn, G. (1966), Über einen Geschlechtunterschied in der mutagenen Wirkung von Trenimon bei der Maus, *Humangenetik 2*, 81–82.

Röhrborn, G., and Berrang, H. (1967), Dominant lethals in young female mice, *Mutation Res. 4*, 231–233.

Röhrborn, G. Kühn, O., Hansmann, I., and Thon K. (1971), Induced chromosome aberrations in early embryogenesis of mice, *Humangenetik 11*, 316–322.

Russell, L. B., and Russell, W. L. (1955), The sensitivity of different stages in oogenesis to the radiation induction of dominant lethals and other changes in the mouse, *in* "Progress in Radiobiology" (J. S. Mitchell, B. E. Holmes, and C. L. Smith, eds.), pp. 187–192, Oliver and Boyd, London.

Russell, L. B., Stelzner, K. F., and Russell, W. L. (1959), Influence of dose rate on radiation effect on fertility of female mice, *Proc. Soc. Exptl. Biol. Med. 102*, 471–479.

Schugt, P. (1928), Untersuchungen über die Wirkung abgestufter Dosen von Roentgenstrahlen verschiedener Wellenlänge auf die Struktur und Funktion der Ovarien, *Strahlentherapie 27*, 603–662.

CHAPTER 33

Insect Chemosterilants as Mutagens

Alexej B. Bořkovec

Insect Chemosterilants Laboratory*
Beltsville, Maryland

I. INTRODUCTION

Chemical compounds that reduce or eliminate the reproductive capacity of an organism are called chemosterilants. In animals, particularly in those that reproduce sexually, normal reproductive processes form a complex network that can be disturbed chemically on several levels: cellular, organic, or behavioral. Since all such disturbances may lead eventually to a decrease or cessation of reproduction, chemosterilants include compounds with different pharmacological properties and different modes of action. Nevertheless, because of the unique practical requirement that these compounds were expected to fulfill, the search for new chemosterilants was oriented primarily to finding and identifying selective types of cytotoxicity. A brief historical summary will clarify the reason for this orientation.

Between 1952 and 1953, scientists of the U.S. Department of Agriculture conducted successful field experiments with a revolutionary method of insect control (Baumhover, 1966). The principle of this new technique was simple: the reproductive capacity of an insect population is inversely related to the ratio between the sterile and fertile members of the popula-

---

*Agricultural Environmental Quality Institute, Agricultural Research Center, Agricultural Research Service, U.S. Department of Agriculture.

tion. If the proportion of sterile insects becomes high enough, the population will decline and under suitable conditions it will die out. Because the principle operates regardless of the size of the population, the appropriate ratio between sterile and fertile insects can be achieved by sterilizing a certain proportion of an existing population or by adding to it a given number of laboratory-reared sterile insects. It was the latter technique that led to the spectacular eradication of the screwworm, *Cochliomyia hominivorax* Coquerel, from the island of Curaçao and later from the southeastern United States. However, although insects could be sterilized by γ-rays and released, for sterilizing naturally occurring insects other techniques had to be developed. E. F. Knipling (1960), the pioneer of this new method of pest control, suggested that chemicals could be used for this purpose, and the search for chemosterilants was initiated. Later it was realized that sterilization by irradiation was not applicable to all insects, and the potential role of chemosterilants was extended to both variants of the sterility-control method.

An ideal chemosterilant should possess a unique set of characteristics (Boŕkovec, 1966), and the most important one is a biological reproductive specificity. Whatever its mode of action, the compound must not interfere with the behavior of the insects, and the longevity and general vigor of the treated insects should remain normal. These restrictions immediately eliminated all general toxicants, and compounds that in some way prevent normal mating. If taken literally, they also would eliminate most other compounds, but, with some allowances for imperfections, a potentially practical chemosterilant can be selected from compounds whose primary effect consists of inactivation of the germ cells or the reproductive organs. The practical considerations that generated this selection trend also determined the design of testing methods. Since a more detailed account of testing will be presented later, it is sufficient to note here that the general procedure for screening candidate chemosterilants is strongly biased toward compounds that are mutagenic or selectively cytotoxic in rapidly growing cellular systems. It is for this reason that the results of chemosterilant screening are of interest to a much wider segment of the scientific community than the relatively small group of pest-control specialists.

The principal question to be decided is whether the testing for chemosterilant activity in insects can provide any information on a compound's mutagenicity. Some 10,000 compounds have been screened in the USDA laboratories since 1960, and many others have been screened by industry and in foreign countries. The total is not known accurately because negative data are almost never reported, but it appears that more compounds were tested as chemosterilants than as mutagens, even if all the mutagenicity testing methods are combined. Therefore, the potential value of chemosterilant testing data is considerable.

II. SCREENING OF CHEMOSTERILANTS

The effects that a chemosterilant can produce in test insects may be characterized as follows:

1. Reduction or cessation of oviposition.
2. Production of eggs that are not viable.
3. Aspermia.
4. Production of sperm that does not form viable zygotes.

Which of these effects may be properly classified as mutagenic is not easy to decide from the commonly available screening data. For example, if a treated female does not produce any eggs, the chemical may have (a) attacked the developing egg and prevented its maturation, (b) attacked the ovarian tissues and destroyed their normal functions, or (c) interfered with the supply of an essential nutrient or a regulatory agent. Each of these possibilities may be further analyzed on a subcellular level. Thus effect (a) may have been the result of an induced mutation within a stem cell in the germanium, or it may have been a specific cytotoxic action affecting the developing egg. Obviously, only the former effect would be properly designated as mutagenic. Even the frequently encountered chemosterilization of the male, in which the insect produces and delivers a normal supplement of motile sperm, is ambiguous from a genetic standpoint. Since zygotic mortality is considered to be the evidence for the presence of dominant lethal mutations in at least one of the gametes, the sperm of a chemosterilized male that produces an inviable zygote should carry such a genetic abnormality. However, an alternate and presumably nongenetic explanation is possible. The chemical may have destroyed the ability of the sperm to penetrate the egg (acrosome reaction), or it may have damaged the intricate mechanism required for pronuclear fusion.

For obvious reasons, the activity of only a very few chemosterilants could be analyzed in such detail, and the standardized screening processes give only scant information on the mechanism by which sterility was induced. As a result, insect chemosterilants cannot be summarily designated as mutagens, and neither can those compounds that failed the screen be cleared as nonmutagenic. A description of a screening technique developed by G. C. LaBrecque (1968) and coworkers in the USDA laboratory in Gainesville, Fla., will illustrate these points. The candidate compound is administered in food to 100 freshly emerged house flies of both sexes. After 6–7 days, the females are allowed to oviposit, and if no eggs are laid the oviposition medium is offered repeatedly in 1–2 day periods. The extent of oviposition is estimated visually, and a random sample of 100 eggs is checked for viability. The unhatched eggs are counted, and the larvae are reared in a larval medium until pupation, when they are counted. The entire process takes about 14 days, but the data give no indication whether males, females, or both sexes were affected. Therefore, when sterility occurs in the treated flies, ten males are removed from the test cage and crossed with ten virgin 4-day-old untreated females. Oviposition, hatch, and pupation are again evaluated, and the results indicate whether males were sterilized.

What value do these results have in regard to mutagenicity of the compounds screened? Negative results, i.e., when there is no difference between the reproductive performance of treated and untreated control flies, would indicate that the compound does not produce dominant lethal mutations under the conditions of the experiment. To the reader, the universal meaning of the latter qualifying restriction should be evident, and no further mention will be made of it. However, a compound that was inactive in the screening may still be mutagenic even in the house fly, if the mutations it produces are nonlethal. Clearly, the screening results can yield no information on this type of mutagenicity. Positive results, on the other hand, indicate a good possibility of mutagenic activity. In spite of the ambiguities mentioned earlier, most of the chemosterilants that were investigated in detail were indeed mutagenic. It is this largely statistical evidence which increases the value of chemosterilant screening data beyond the realm of economic entomology. Conceivably, research on chemosterilants could profit more from the data on all types of mutagenicity than *vice versa*, but at present the number of compounds screened for mutagenicity is small compared to the number of screened candidate chemosterilants.

Only estimates of the numbers of compounds tested as chemosterilants can be offered. Since 1955, when insect sterility acquired a certain economic potential, close to 10,000 compounds have been screened in various USDA laboratories, primarily in house flies, screwworms, Mexican fruit flies (*Anastrepha ludens* Loew), and boll weevils (*Anthonomus grandis* Boheman). Some 600 of these compounds had more or less pronounced streilizing effects, and many were further tested in over 100 species of insects and other organisms. Possibly thousands of additional compounds were screened in other laboratories here and abroad, but an exact count of the active and inactive compounds has not been made. By the end of 1972, more than 1000 insect chemosterilants were described in the scientific literature. This number is more than twice that listed in a survey published by Bořkovec (1966).

III. CLASSIFICATION OF CHEMOSTERILANTS

A most striking parallel between chemosterilants and mutagens becomes apparent when the types of active compounds in each category are compared. A strong similarity also exists between chemosterilants and carcinostatic compounds; the relationship to carcinogens is less pronounced. No effort will be made here to discuss the classification and specific characteristics of mutagens; the reader is referred to Chapters 1 and 3 of the present work or to other specialized literature.

Historically, the search for chemosterilants derived its first clues from the work on chemotherapy of cancer (Bořkovec, 1962). Prior to 1955, numerous publications described the use of antimetabolites and alkylating agents in reducing fertility in insects, and the preferential biological activity of these compounds in rapidly proliferating cells was well known. Since most adult insects contain few such cell systems except the gonads, antitumor compounds appeared to be an ideal source of model chemosterilants. Even now, valuable leads are occasionally found among new antitumor agents. The Official National Cancer Institute list of effective antineoplastic drugs (Wood, 1971) contains alkylating agents, antimetabolites, hormonal agents, and miscellaneous compounds. A similar classification, presented in Table 1 in a slightly expanded form, has been used for classifying chemosterilants.

The first two classes are well known to students of mutagenesis and require no further discussion. Suffice it here to say that the aziridines (la), and more recently also certain alkanesulfonates (lc), are still the most promising compounds for sterilizing captive insects. The antimetabolites, on the other hand, are of limited interest because of their frequent lack of sterilizing effects in male insects. The third class, hormonal agents, includes an unusual group of insect hormones and their analogues (3a) that are not known to have any antineoplastic or mutagenic properties. These terpenoid compounds, however, were discovered only recently, and their toxicology and other biological activities are still being studied.

It is in the category of miscellaneous compounds where the relationship between sterilizing activity and mutagenicity requires testing and clarification. Prior to the discovery (Chang et al., 1964) of the sterilizing activity of nonalkylating s-triazines (4a), this class of widely used compounds was not known to contain distinct mutagens. Nevertheless, the mutagenicity of hemel (hexamethylmelamine) has been reported (Beneš and Šrám, 1969; Chang and Klassen, 1968), and thus the entire class of s-triazines should be now investigated. Fortunately, at least as far as chemosterilization is concerned, the economically most important s-triazine herbicides had very low or no sterilizing activity (Bořkovec et al., 1967), and the structure–activity relationship in this type of compound appeared to be highly specific.

Similar chemosterilant (Chang et al., 1964) and mutagenic (Chang and Klassen, 1968; Beneš and Šrám, 1969; Ninan and Wilson, 1969;

TABLE 1. Classification of Insect Chemosterilants

Class of compounds	Representative structure	General references[a]
1. Alkylating agents a. Aziridines	TEPA	Bořkovec (1969), Bořkovec *et al.* (1966, 1968), Chang *et al.* (1970)
b. Nitrogen mustards	$CH_3N{<}^{CH_2CH_2Cl}_{CH_2CH_2Cl}$ Mechlorethamine	Bořkovec (1969)
c. Sulfonic acid esters	$CH_3SO_2O(CH_2)_4OSO_2CH_3$ Busulfan	Bořkovec (1969), Crystal (1968), Klassen *et al.* (1968)
d. Other alkylating agents	$ICH_2CH(OH)CH_2OH$ 3-Iodo-1,2-propanediol	DeMilo and Crystal (1972)
2. Antimetabolites a. Purine and pyrimidine analogues	5-Fluorouracil	Cline (1968), Jalil and Morrison (1969) Řežábová (1968), Řežábová and Landa (1967)
b. Folic acid analogues	Methortrexate	Akov (1967), Řežábová (1968), Sugai and Ashoush (1970)
c. Other antimetabolites	$CH_3CH_2SCH_2CH_2CHCOOH$ $\quad\quad\quad\quad\quad\;\; NH_2$ Ethionine	Guerra (1970), Hegdekar (1970)
3. Hormonal agents a. Juvenile hormone analogues	Methyl 3,7,11-trimethyl-7,11-dichloro-2-dodecenoate	Masner *et al.* (1970)

TABLE 1. Cont'd

Class of compounds	Representative structure	General references[a]
b. Steroids	 Ecdysone	Řežábová *et al.* (1968) Robbins *et al.* (1968)
4. Miscellaneous compounds a. Nonalkylating *s*-triazines	 Hemel	Bořkovec *et al.* (1972), Bořkovec and De- Milo (1967), LaBrecque *et al.* (1968)
b. Nonalkylating phosphoramides	 Hempa	Klassen *et al.* (1968), Terry and Bořkovec (1967) Terry and Crystal (1972)
c. Tin derivatives	 Triphenyltin hydroxide	Ascher *et al.* (1968)
d. Boron derivatives	 Benzeneboronic acid	Bořkovec *et al.* (1969), Settepani *et al.* (1969, 1970)
e. Dithiazoles	 3,5-Bis(dimethylamino)-1,2,4-dithiazolium chloride	Chang *et al.* (1972), Oliver *et al.* (1972)

TABLE 1. Cont'd

Class of compounds	Representative structure	General references[a]
f. Alkaloids	Monocrotaline	Guerra (1972), Klassen *et al.* (1968) Wicht and Hays (1967)
g. Antibiotics	Anthramycin	Bořkovec *et al.* (1971), Horwitz *et al.* (1971)

[a]For references prior to 1966, see Bořkovec (1966).

Šrám *et al.*, 1970; Šrám, 1972) activity was found in the nonalkylating phosphorus amides. Alkylating phosphorus amides and *s*-triazines, e.g., TEPA [tris(1-aziridinyl)phosphine oxide] and tretamine [2,4,6-tris-(1-aziridinyl)-*s*-triazine], have long been known as strong mutagens and more recently also as chemosterilants. However, in analogy with other aziridinyl compounds, it was believed that their activity was owing to the aziridinyl group and that the remaining portion of the molecule, i.e., the phosphorus or the triazinyl moiety, was only a biologically inactive carrier. This hypothesis may still hold true, but the chemosterilant and mutagenic activities of hempa (hexamethylphosphoric triamide) and of the already mentioned hemel indicate that even a seemingly innocuous functional group when attached to a suitable carrier may give rise to unexpected biological activity.

Most of the chemosterilants in classes 4a, 4b, and 4d have not been tested for mutagenic activity by any of the standardized independent procedures discussed in Volumes 1 and 2. Consequently, the only structure–activity correlations are those derived from chemosterilant research. One of the most unusual entities that figures prominently in the mentioned classes of sterilants is the dimethylamino group. Hemel and other dimethylaminophosphine oxides and sulfides, and also several (dimethylamino) dithiazolium compounds (4e), all contain this group. Examination of the metabolism of hemel (Chang *et al.*, 1968) and hempa (Chang *et al.*, 1967) revealed a pathway that included hydroxymethyl intermediates (Terry and Bořkovec, 1970) that may conceivably function as alkylating agents. These findings led to a hypothesis that the dimethylamino sterilants required metabolic activation and conversion to alkylating agents, which then functioned in the typical way known for compounds in class 1. However, the metabolism of dithiazoles, in which the dimethylamino group also appears to have a conspicuous role, has not been investigated, and chemical considerations do not support a similar activation hypothesis for this class of compounds.

In class 4c, the chemosterilant activity is fairly restricted to triphenyltin derivatives, and no mutagenicity data are known for such compounds. There is some controversy regarding the activity of tin chemosterilants in male insects, but in general these compounds are more effective in females than in males. Because a normal function of ovaries depends on a balanced interplay of nutritive, hormonal, and possibly other factors that do not seem to be of similar importance in male gametogenesis, the relationship between sterilizing and mutagenic activity in female chemosterilants is more complex than that in male sterilants. Also, the practical potential of now known tin chemosterilants is limited, and no detailed studies on their metabolism and mode of action have been conducted. Similar considerations apply to boron derivatives (4d), which also are active predominantly in female insects. Nevertheless, because the activity of certain boron compounds has been traced to boric acid (Settepani *et al.*, 1969), a widely used commercial chemical, the implication of possible mutagenic properties in this type of compounds deserves further study.

From a chemical standpoint, the compounds in classes 4f and 4g constitute an extremely difficult problem for structure–activity studies. Some of them, e.g., the pyrrolizidine alkaloids, have been investigated in great detail (Fishbein *et al.*, 1970, and references therein), and their mutagenic and sterilizing activity is well documented. Others, particularly those that were discovered only recently, are still under investigation, and their full biological evaluation will have to await further development.

IV. CONCLUSION

The study of insect chemosterilants is oriented toward a possible utilization of these compounds as pest-control agents. An intregral part of any such endeavor is the investigation of toxicological properties of materials that, if used in practice, could occur as residues or environmental contaminants. It should be apparent from the foregoing discussion that the relationship between chemosterilants and mutagens is not direct and simple and that only a fraction of known chemosterilants were tested and found mutagenic in independent experiments. Similarly, the mutagenic basis for chemosterilant activity has also been demonstrated in only a handful of cases. Nevertheless, the safety requirements of the use of chemosterilants make it mandatory to look beyond the common toxicological standards and tests, and a mutagenic potential of any practical chemosterilant must be evaluated whenever the danger of contamination presents itself. Fortunately, this need has been long recognized by most scientists working with chemosterilants, and particularly the USDA research in this field was and is oriented not only to discovering new compounds but also to finding safe procedures for handling and using any chemosterilants. It may be noted in this connection that no insect chemosterilant has as yet been approved or recommended for use by any Agency of the U.S. Government.

Although the question of safety of chemosterilants in light of the growing awareness of mutagenic hazards has received much attention recently, the possible impact that chemosterilant research may have on the study of mutagenicity has been largely neglected (Bořkovec, 1970). It was mentioned earlier that several new types of mutagens were discovered via their first-found sterilizing activity. Undoubtedly, additional chemosterilants will be tested and their mutagenicity characterized. The existing list of chemosterilants (Bořkovec, 1966) is now outdated, and since the inclusion of a current list of these compounds would be beyond the scope of this chapter, only scattered references in entomological and chemical literature show the full range of presently known materials. Even less accessible are the data on thousands of candidate compounds that were inactive. Closer coordination of research in both areas is clearly needed, and if some of the gaps created by the highly specialized testing for chemosterilant activity could be filled by geneticists and other workers interested in mutagenesis, at least some of the relationships between the two biological activities may be resolved.

V. REFERENCES

Akov, S. (1967), Effect of folic acid antagonists on larval development and egg production in *Aedes aegypti*, *J. Insect Physiol. 13*, 913–923.

Ascher, K. R. S., Meisner, J., and Nissim, S. (1968), The effect of fentins on the fertility of the male house fly, *World Rev. Pest Control 7*, 84–96.

Baumhover, A. H. (1966), Eradication of the screwworm fly, *J. Am. Med. Ass. 196*, 240–248.

Beneš, V., and Šrám, R. (1969), Mutagenic activity of some pesticides in *Drosophila melanogaster*, *Industr. Med. 38*, 442–444.

Bořkovec, A. B. (1962), Sexual sterilization of insects by chemicals, *Science 137*, 1034–1037.

Bořkovec, A. B. (1966), "Insect Chemosterilants," 143 pp., Interscience Publishers, New York.

Bořkovec, A. B. (1969), Alkylating agents as insect chemosterilants, *Ann. N. Y. Acad. Sci. 163* (Art. 2), 860–868.

Bořkovec, A. B. (1970), Chemical mutagenesis and the testing of insect chemosterilants, *EMS Newsletter 1(3)*, 10–11.

Bořkovec, A. B., and DeMilo, A. B. (1967), Insect chemosterilants. V. Derivatives of melamines, *J. Med. Chem. 10*, 457–461.

Bořkovec, A. B., Woods, C. W., and Brown, R. T. (1966), Insect chemosterilants. 3. 1-Aziridinylphosphine oxides, *J. Med. Chem. 9*, 522–526.

Bořkovec, A. B., LaBrecque, G. C., and DeMilo, A. B. (1967), s-Triazine herbicides as chemosterilants of house flies, *J. Econ. Entomol. 60*, 893–894.

Bořkovec, A. B., Fye, R. L., and LaBrecque, G. C. (1968), Aziridinyl chemosterilants for house flies, *USDA ARS (Ser.) 33-129*, 60 pp.

Bořkovec, A. B., Settepani, J. A., LaBrecque, G. C., and Fye, R. L. (1969), Boron compounds as chemosterilants for house flies, *J. Econ. Entomol. 62*, 1472–1480.

Bořkovec, A. B., Chang, S. C., and Horwitz, S. B. (1971), Chemosterilization of house flies with anthramycin methyl ether, *J. Econ. Entomol. 64*, 983–984.

Bořkovec, A. B., DeMilo, A. B., and Fye, R. L. (1972), Triazinyl chemosterilants for the house fly: 2, 4-Diamino-s-triazines, *J. Econ. Entomol. 65*, 69–73.

Chang, S. C., Terry, P. H., and Bořkovec, A. B. (1964), Insect chemosterilants with low toxicity for mammals, *Science 144*, 57–68.

Chang, S. C., Terry, P. H., Woods, C. W., and Bořkovec, A. B. (1967), Metabolism of hempa uniformly labeled with C^{14} in male house flies, *J. Econ. Entomol. 60*, 1623–1631.

Chang, S. C., DeMilo, A. B., Woods, C. W., and Bořkovec, A. B. (1968), Metabolism of C^{14}-labeled hemel in male house flies, *J. Econ. Entomol. 61*, 1357–1365.

Chang, S. C., Woods, C. W., and Bořkovec, A. B. (1970), Sterilizing activity of bis (1-aziridinyl) phosphine oxides and sulfides in male house flies, *J. Econ. Entomol. 63*, 1744–1746.

Chang, S. C., Oliver, J. E., Brown, R. T., and Bořkovec, A. B. (1972), Chemosterilization of male house flies with dithiazolium compounds and with tetramethyldithiobiuret, *J. Econ. Entomol. 65*, 390–392.

Chang, T. H., and Klassen, W. (1968), Comparative effects of tretamine, tepa, apholate and their structural analogs on human chromosomes *in vitro*, *Chromosoma (Berlin) 24*, 314–323.

Cline, R. E. (1968), Evaluation of chemosterilant damage to the testes of the housefly, *Musca domestica*, by microscopic observation and by measurement of the uptake of ^{14}C-compounds, *J. Insect Physiol. 14*, 945–953.

Crystal, M. M. (1968), Sulfonic acid esters as chemosterilants of screwworm flies with particular reference to methanediol dimethanesulfonate, *J. Econ. Entomol. 61*, 446–449.

DeMilo, A. B., and Crystal, M. M. (1972), Chemosterilants against screwworm flies. 3, *J. Econ. Entomol. 65*, 594–595.

Fishbein, L., Flamm, W. G., and Falk, H. L. (1970) "Chemical Mutagens," pp. 176–179, Academic Press, New York.

Guerra, A. A. (1970), Effect of biologically active substances in the diet on development and reproduction of *Heliothis* spp., *J. Econ. Entomol. 63*, 1518–1521.

Guerra, A. A. (1972), Sterility induced in tobacco budworms by combination of reserpine and gamma irradiation affected by age and sex of pupae, *J. Econ. Entomol. 65*, 1282–1283.

Hegdekar, B. M. (1970), Amino acid analogues as inhibitors of insect reproduction, *J. Econ. Entomol. 63*, 1950–1956.

Horwitz, S. B., Chang, S. C., Grollman, A. P., and Bořkovec, A. B. (1971), Chemosterilant action of anthramycin: A proposed mechanism, *Science 174*, 159–161.

Jalil, M., and Morrison, P. E. (1969), Chemosterilization of the two-spotted spider mite. I. Effect of chemosterilants on the biology of the mite, *J. Econ. Entomol. 62*, 393–400.

Klassen, W., Norland, J. F., and Bořkovec, A. B. (1968), Potential chemosterilants for boll weevils, *J. Econ. Entomol. 61*, 401–407.

Knipling, E. F. (1960), Use of insects for their own destruction, *J. Econ. Entomol. 53*, 415–420.

LaBrecque, G. C. (1968), Laboratory procedures, in "Principles of Insect Chemosterilization" (G. C. LaBrecque and C. S. Smith, eds.), pp. 41–98, Appleton-Century-Crofts, New York.

LaBrecque, G. C., Fye, R. L., DeMilo, A. B., and Bořkovec, A. B. (1968), Substituted melamines as chemosterilants of house flies, *J. Econ. Entomol. 61*, 1621–1632.

Masner, P., Sláma, K., Žd'árek, J., and Landa, V. (1970), Natural and synthetic materials with insect hormone activity. X. A method of sexually spread insect sterility, *J. Econ. Entomol. 63*, 706–710.

Ninan, T., and Wilson, G. B. (1969), Chromosome breakage by ethylenimines and related compounds, *Genetics 40*, 103–119.

Oliver, J. E., Chang, S. C., Brown, R. T., Stokes, J. B., and Bořkovec, A. B. (1972), Insect chemosterilants. 11. Substituted 3,5-diamino-1,2,4-dithiazolium salts and related compounds, *J. Med. Chem. 15*, 315–320.

Řežábová, B. (1968), Changes in the metabolism of nucleic acids in the ovaries of the house fly, *Musca domestica*, after application of chemosterilants, *Acta Entomol. Bohemoslov. 65*, 331–340.

Řežábová, B., and Landa, V. (1967), Effect of 6-azauridine on the development of the ovaries of the house fly, *Musca domestica* L. (Diptera), *Acta Entomol. Bohemoslov. 64*, 344–351.

Řežábová, B., Hora, J., Landa, V., Černý, V., and Šorm (1968), On steroids. CXIII. Sterilizing effects of some 6-ketosteriods on housefly *(Musca domestica* L.), Steroids *11*, 475–496.

Robbins, W. E., Kaplanis, J. N., Thompson, M. J., Shortino, T. J., Cohen, C. F., and Joyner, S. C. (1968), Ecdysones and analogs: Effects on development and reproduction of insects, *Science 161*, 1158–1160.

Settepani, J. A., Crystal, M. M., and Bořkovec, A. B. (1969), Boron chemosterilants against screw-worm flies: Structure–activity relationship, *J. Econ. Entomol. 62*, 375–383.

Settepani, J. A., Stokes, J. B., and Bořkovec, A. B. (1970), Insect chemosterilants. VIII. Boron compounds, *J. Med. Chem. 13*, 128–131.

Šrám, R. J. (1972), The differences in the spectra of genetic changes in *Drosophila melanogaster* induced by chemosterilants tepa and hempa, *Folia Biol. (Praha) 18*, 139–148.

Šrám, R. J., Beneš, V., and Zudová, Z. (1970), Induction of dominant lethals in mice by tepa and hempa, *Folia Biol. (Praha) 16*, 407–416.

Sugai, E., and Ashoush, I. (1970), Male sterility in the silkworm, *Bombyx mori* L., induced by aminopterin (Lepidoptera: Bombycidae), *Appl. Entomol. Zool. 5*, 202–207.

Terry, P. H., and Bořkovec, A. B. (1967), Insect chemosterilants. IV. Phosphoramides, *J. Med. Chem. 10*, 118–119.

Terry, P. H., and Bořkovec, A. B. (1970), Insect chemosterilants. IX. N-(Hydroxymethyl)-N,N',N',N'',N''–pentamethylphosphoric triamide, *J. Med. Chem. 13*, 782–783.

Terry, P. H., and Crystal, M. M. (1972), Chemosterilants against screwworm flies. 2, *J. Econ. Entomol. 65*, 307–310.

Wicht, M. C., Jr., and Hays, S. B. (1967), Effect of reserpine on reproduction of the house fly, *J. Econ. Entomol. 60*, 36–38.

Wood, H. B., Jr. (1971), Selection of agents for the tumor screen, *Cancer Chemotherap. Rep. (Part 3) 2*, 9–22.

CHAPTER 34

The Literature of Chemical Mutagenesis

John S. Wassom

Biology Division
Oak Ridge National Laboratory
Oak Ridge, Tennessee

I. INTRODUCTION

A. Restatement of an Old Problem

Keeping aware of new developments in science today is not an easy task. Earlier, this was not a problem, because information could be obtained through communicating with colleagues and keeping abreast with published reports in prominent scientific journals. As science has advanced, these methods of acquiring information have become insufficient. The cause for this change has been the large volume of reports which reflect the technological advancements made in old and newly discovered research areas. This information increase has made it necessary to use our technical skills to develop adequate control measures to prohibit scientific information from becoming a mass of inaccessible, irretrievable, and duplicated work. The implementation of automated efforts in information control requires that scientists and other interested individuals stay familiar with the literature of their respective disciplines.

B. Chemical Mutagenesis and Its Literature

Even though chemical mutagenesis is a relatively new research field,

TABLE 1. Multidisciplinary Procurement Resources for Chemical Mutagenesis Information

Data base	Producer	Scope	Coverage	Services
Biological Abstracts	Biosciences Information Service, Philadelphia, Pennsylvania	Life sciences	8000 source publications from 100 countries	1. A subject-oriented author-indexed collection of abstracts published twice monthly under the title *Biological Abstracts* 2. Computerized searches are available either from the producer or from institutions which have purchased these tapes for use in their search services
BioResearch Index	Biosciences Information Service, Philadelphia, Pennsylvania	Life sciences	More than 100,000 articles from the following sources are reported annually: institutional reports, bibliographies, letters, notes, preliminary reports, reviews, government reprints, semipopular journals, symposia, trade journals	1. Issues a monthly publication containing bibliographical information and some abstracts 2. Computerized searches are available either from the producer or from institutions which have purchased these tapes for use in their search services
Chemical Abstracts	American Chemical Society, Columbus, Ohio	World chemical literature	Approximately 1000 primary journals	1. Issues a weekly collection of abstracted information which has been indexed according to subject keyword, numerical patent, patent concordance, and author

				2. Magnetic tapes are available from producer for sale to customers for searching; searches are also available from institutions which have purchased these tapes for use in their search services
Chemical-Biological Activities	Chemical Abstracts, Research Division of the American Chemical Society, Columbus, Ohio	Effects of organic chemicals on biological systems	Approximately 600 life science and chemical journals	1. Produces a monthly document containing a bibliographical-abstract section, a keyword in context section, a molecular formula index, cross-reference indexed; structural formulas appear with abstracts 2. Magnetic tapes are available from producer for sale to customers for searching. Searches are also available from institutions which have purchased these tapes for use in their search services.
Chemical Titles	American Chemical Society, Columbus, Ohio	Current awareness of titles from journals of pure and applied chemistry	700 journals	Issues a biweekly publication with author and KWIC indexes
Index Medicus	National Library of Medicine, Washington, D. C.	Biomedical literature	Approximately 2300 sources of periodical literature	Issues a monthly subject- and author indexed bibliography

TABLE 1 Cont'd

Data base	Producer	Scope	Coverage	Services
MEDLine (Medical Literature Analysis and Retrieval System on Line)	National Library of Medicine, Washington, D. C.	Biomedical literature	Approximately 1260 sources of periodical literature	Provides computer searches of the tape version of Index Medicus subject materials
Current Contents	Institute for Scientific Information, Philadelphia, Pennsylvania	Life sciences	Approximately 1000 journals	Issues a weekly collection of the tables of contents of screened journals
ASCA (Automated Science Citation Alert)	Institute for Scientific Information, Philadelphia, Pennsylvania	Life sciences	Approximately 1000 journals	Provides computerized search of tables of contents of key journals by journal name and/or title keywords
Genetics Abstracts	Information Retrieval Limited, London, England	Literature of genetics and related disciplines	3000 journals	Issues a monthly collection of subject-indexed abstracts
Carcinogenesis Abstracts	National Cancer Institute, Bethesda, Maryland	Literature reporting on carcinogenesis research	Not available	Issues a monthly collection of subject-indexed abstracts
Cancer Chemotherapy Abstracts	National Cancer Institute, Bethesda, Maryland	Literature reporting on carcinogenesis research	Not available	Issues a monthly collection of subject-indexed abstracts
Biological and Agricultural Index	The T. H. Wilson Company, the Bronx, New York	Literature in the field of biology, agriculture, and related sciences	Approximately 150 English-language periodicals	Issues a monthly collection of subject-indexed abstracts

NSA (Nuclear Science Abstracts)	Atomic Energy Commission's Division of Technical Information, Oak Ridge, Tennessee	International nuclear scientific literature	Approximately 2400 journals	1. Issues a semimonthly collection of author- and subject-indexed abstracts 2. Computerized searches of this abstracted information are available from the producer
Excerpta Medica	Excerpta Medica Foundation, Amsterdam, the Netherlands	World biomedical literature	Approximately 3400 journals	Produces a monthly collection of subject- and author-indexed abstracts

it has already acquired its own unique and rapidly expanding literature, complete with all its problems. The risks of genetic damage to man from chemicals render urgency to an understanding of this literature, since extensive retrospective and current literature searching will be basic tools in efforts to minimize exposure to hazardous substances. This chapter is a comprehensive review of this literature field and is presented in the hope that it will help make the recorded information of this science easier to find and use (see Table 1 for a listing of procurement resources available for chemical mutagenesis information).

II. THE STATE OF CHEMICAL MUTAGENESIS LITERATURE

The evolution of publications containing chemical mutagenesis information and their origin as a literature field have logically paralleled the growth and development of research on the genetic effects produced by chemicals. Readers who wish to study the conditions influencing the growth patterns of this new literature field may do so by reviewing the history of chemical mutagenesis as a science. Dr. Charlotte Auerbach has in this volume traced this historical development for review, and many of the events surrounding the publication of many of the classic works with chemical mutagens can be appreciated from her candid discussion.

During the last 10 years, the relationship between man and his environment has become a demanding issue in all sciences. In our case, man's exposure to the many chemicals present in our surroundings prompted some scientists to suspect that some of these substances could be detrimental to man's genetic health. The expression of these fears began appearing in the literature during the late 1960s. This environmental influence has created a new climate for research with its demands for testing priorities, new assay procedures, and population-monitoring techniques. These areas have added another dimension to the extensive knowledge that has accumulated in this literature field since the early days when research efforts were dominated by pure rather than applied research. Today, environmentally oriented research has contributed significantly to the enormous growth of chemical mutagenesis literature. This can be seen by a review of the number of publications produced each year. For instance, from nominal yearly increases prior to 1960, the rate of publications increased almost twofold during the decade of the 1960s over that during the late 1940s and 1950s. Since 1968, published material has grown at a rate of a 200–500 article increase per year (Fig. 1). By using these figures as guides, we estimate that approximately 2500 papers were published during 1972.

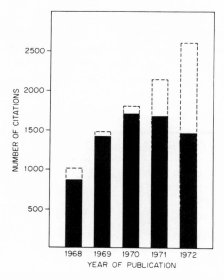

FIGURE 1. Projected yearly increase in chemical mutagenesis literature. The solid portion of the block represents articles presently in EMIC's data base; the broken-line extension represents the approximate number of articles EMIC projects that it will accumulate.

As for the future, there should be a continued rise in published material commensurate with the level of experimentation and interest this field occupies in the years ahead. Also, the publication of negative data heretofore not published with any regularity promises to be a factor contributing to the multiple growth expectancy of this literature field.

III. SOURCES AND TYPES OF CHEMICAL MUTAGENESIS LITERATURE

The word "literature" is defined in this study as that body of writings pertaining to a particular subject, which here is chemical mutagenesis. Writings on any scientific subject may appear in several different forms, such as review articles, symposium proceedings, and conference abstracts, but for the most part writings on chemical mutagenesis appear in journals as papers giving detailed descriptions of research results. There are a wide range of sources to which an author may submit his writings for publication. As science has expanded in scope, certain publication sources have evolved which accept and publish articles along strict disciplinary lines. Most publications on chemical mutagenesis are presently found in sources publishing only genetic information.

From studies made on publication sources for chemical mutagenesis, we have learned that writings about chemical mutagens can presently be found in approximately 700 different sources and in many different forms. To illustrate, we have compiled information on 6094 citations regarding their source of origin and published form. Table 2 is a listing of the various forms in which these citations appeared, and Table 3 lists publi-

TABLE 2. The Literature Forms of Chemical Mutagenesis

Type of published information	Number of citations
Presentation of experimental data as original articles in scientific journals	4957
Abstracts of meetings	576
Review articles	137
Laboratory reports	115
Symposium proceedings	108
Brief notes in scientific newsletters or bulletins	69
Letters to the Editor	50
Feature articles in scientific publications	27
Book chapters	26
Editorial comments	18
Articles proposing hypotheses	8
Book reviews	3

TABLE 3. High-Frequency Sources for Chemical Mutagenesis Information

Publication source	Number of citations
Mutation Research	410
Genetika (USSR)	251
Genetics (US)	248
Nature	164
Journal of Bacteriology	136
Cancer Research	130
Molecular and General Genetics	127
Science	106
Environmental Mutagen Society Newsletter	108
Proceedings of the National Academy of Science (US)	101
Biochimica et Biophysica Acta	97
Hereditas	97
Experimental Cell Research	94
Comptes Rendus Hebdomadaires des Seances de l'Academie des Sciences. Series D. Sciences Naturelles (Paris)	72
Journal of Molecular Biology	67
Arabadopsis Information Service	65
Japanese Journal of Genetics	62
Experientia	60

TABLE 3. Cont'd

Publication source	Number of citations
Doklady Biological Sciences (USSR) [English translation of *Doklady Akademii Nauk SSSR*]	57
International Cancer Congress Abstracts	55
Canadian Journal of Genetics and Cytology	55
Nucleus (Calcutta)	52
Biochemical and Biophysical Research Communications	51
Journal of Cell Biology	50
Tsitologiya	49
Humangenetik	49
Radiation Research	47
Virology	46
Journal of Virology	42
Radiation Botany	41
Chemico-Biological Interactions	40
Biochemical Journal	39
Biochemical Pharmacology	39
Journal of General Microbiology	38
Journal of the National Cancer Institute	37
Microbial Genetics Bulletin	36
Lancet	35
Tsitologiya i Genetika	35
Cytologia	35
Proceedings of the Society of Experimental Biology and Medicine	33
Chromosoma	31
Toxicology and Applied Pharmacology	31
Proceedings of the American Association for Cancer Research	31
Zeitschrift für Verebungslehre [title changed to *Molecular and General Genetics*]	30
Biochemistry	30
Cold Spring Harbor Symposia on Quantitative Biology	30
Naturwissenschaften	29
Bulletin of Experimental Biology and Medicine (USSR) [English translation of *Byulleten' Eksperimental'noi Biologii i Meditsiny*]	29
Zeitschrift für Krebsforschung	29
Federation Proceedings	28
Comptes Rendus des Seances de la Société de Biologie et de ses Filiales	27
Zeitschrift für Naturforschung	27
Gann	27
Annals of the New York Academy of Sciences	26
Carylogia	26
Biopolymers	26
Current Science	25
Genetical Research	25
Indian Journal of Genetics and Plant Breeding	25
Folia Microbiologica	25
Journal of Heredity	24
Journal of Reproduction and Fertility	24

TABLE 3. Cont'd

Publication source	Number of citations
International Journal of Radiation Biology	23
European Journal of Cancer	21
Proceedings of the Symposium on Radiation and Radiomimetic Substances in Mutation Breeding (1969)	21
Abhandlungen der Deutschen Adademie der Wissenschaften zu Berlin Klasse für Medicin	21
International Atomic Energy Agency Technical Reports Series	20
Abstracts of the Eleventh International Botanical Congress	18
Drosophila Information Service	18
Microbiology (USSR) [English translation of *Mikrobiologiya*]	18
Canadian Journal of Microbiology	18
Collection of Papers Presented at the Annual Symposium of Fundamental Cancer Research	16
Neoplasma Bratislava	16
British Medical Journal	16
Biologia Plantarum	15
Radiobiology (USSR) [English translation of *Radiobiologiya*]	15
Molecular Biology (USSR) [English translation of *Molekulyarnaya Biologiya*]	15
Scientia (Milano) Rivista Internazionale di Sintesi Scientifica	14
Journal of Cellular Physiology	14
Journal of Medicinal Chemistry	14
New England Journal of Medicine	12
Archiv für Mikrobiologie	12
European Journal of Biochemistry	12
Antibiotiki	11
FEBS [Federation of European Biochemical Societies]	11
Applied Microbiology	11
Nature New Biology	11
Acta Virologica	11
Proceedings of the Indian Science Congress	11
Journal of the American Medical Association	11
Deutsche Medizinische Wochenschrift	10
Zeitschrift für Allgemeine Mikrobiologie	10
Archiv für die Gesamte Virusforschung	10
Archiv für Pharmakologie und Experimentelle Pathologie (Naunyn-Schmiedebergs)	10
Argonne National Laboratory Annual Report	10
Izvestiya Akademii Nauk SSSR Seriya Biologicheskaya	10
Journal of Biological Chemistry	10
Food and Cosmetics Toxicology	10
Arzneimittel-Forschung	10
Journal of General Virology	10
Teratology	10
Symposia of the Czechoslovak Academy of Sciences	10
Biophysics (USSR) [English translation of *Biofizika*]	10

TABLE 3. Cont'd

Publication source	Number of citations
American Naturalist	10
American Journal of Botany	10
Molecular Pharmacology	10
Journal of Genetics	10
Chemical and Pharmaceutical Bulletin	10
Theoretical and Applied Genetics	10

cation sources which carried at least ten of these citations. All the 700 sources carrying these references have not been listed, because of space limitations, but the list provided does reveal names of publications that do carry a high frequency of articles relating to chemical mutagenesis.

IV. ORGANIZATION OF CHEMICAL MUTAGENESIS INFORMATION

In the introduction to this chapter, we stated that there was a need for organizing the literature in all areas of science because of the phenomenal increase in information. The development of chemical mutagenesis as a literature field was also reviewed by paralleling its growth with that of chemical mutagenesis as a science. As a distinct literature area, we learned that chemical mutagenesis information is available in many different forms and from hundreds of sources. In the following section, we wish to explore how the problems of organizing this information field are being dealt with.

A. General Methods

Prior to 1969, the transfer of chemical mutagenesis information from its origin to the user was made possible by communicating with colleagues, monitoring articles in key journals, or using some of the large abstracting services. The inadequacies of searching the literature by using these first two methods can be explained on the basis of volume. The third process, of using large abstracting services, will now be considered. These abstracting facilities are multidisciplinary in nature, and many employ automated methods as aids in the retrieval and transfer of information. This method of information control was instituted as man's capabilities of keeping track of data diminished. Today, there are several of these facilities which collect, process, and disseminate information of importance to chemical mutagenesis. These activities are listed in Table 1.

Services offered by these facilities are normally twofold. They produce an indexed collection of abstracts which are published at various intervals and are available by subscription. Also, computer searches can be implemented on many of these data holdings and are available on a continuing- or single-search basis. Some of these clearinghouses have been in existence for a long time, while others are of more recent origin, and all have become quite proficient considering the magnitude of their operations. The use of these sources individually, however, for comprehensive coverage of the chemical mutagenesis literature is not possible. As the volume of literature has increased, the degree of coverage of these facilities has diminished. Presently, in order to ensure adequate coverage of the chemical mutagenesis literature, the services of all of these broad-scoped facilities must be employed along with the standard methods of personal contact and manual journal monitoring.

B. Specialized Methods

In 1969, some of the members of the newly formed Environmental Mutagen Society felt that since access to chemical mutagenesis information had exceeded individual capabilities, there was a need for a centralized effort to collect and organize this information. In September 1969, the Environmental Mutagen Information Center (EMIC) was created to begin this undertaking. The Center was commissioned to collect, store, and disseminate information pertaining to chemical mutagenesis. Presently, EMIC has 8500 data holdings, most of which were published during the years 1969–1972. The future promises a steady increase in this number.

In order to better acquaint you with EMIC, I have outlined its operation in Fig. 2.

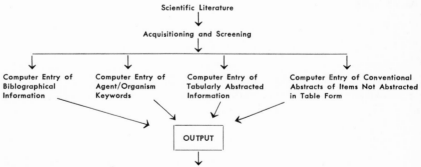

FIGURE 2. EMIC's processing format for chemical mutagenesis information.

(Documentalist)	(Input Team)	(Date of Preparation)
ESV/MS	EMIC	7-26-72

$<$ Literature Type $>$ Journal $<$ Publication Date $>$ 1972 $<$ Header $>$ 012070
$<$ Author $>$ Rodriguez Murcia, C.; Arroyo Nombela, J.J.
$<$ Title $>$ Cytological Aberrations Produced by Methotrexate in Mouse Ascites Tumors
$<$ Publication Description $>$ Mutation Research 14: 405–412
$<$ Selection Source $>$ MEDLARS
$<$ Test Object $>$ Mammal, Mouse Ascites Tumor
$<$ TAXON $>$ MUS, Strain NMR I
$<$ Agent keyword $>$ 4-Amino-10-methylpteroylglutamic acid (Methotrexate)

FIGURE 3. Example of EMIC's entry of bibliographical and agent/organism data.

EMIC employs all the services listed in Table 1 as well as the afore-mentioned standard methods of data procurement to obtain information. The data received from these screening methods are scanned, and references are selected for entry into EMIC's data base. Information which has been earmarked for EMIC's use is further verified by procuring and reviewing a hard copy of the citation whenever possible. After verification, bibliographical data and agent/organism keywords are recorded and sent to the computer. Figure 3 shows examples of these types of entries. Further processing of a document depends on the type of literature at hand. For instance, all items in which experimental results of original work are presented are tabularly abstracted. Documents not indexed in tabular form, such as review articles, meeting abstracts, and editorial comments, have concise written abstracts prepared on them along with appropriate descriptive terms.

All these methods of information processing except tabular abstracting are used quite commonly today and are relatively self-explanatory as to their function. Tabular abstracting is a new approach for use in the computer manipulation of mutagenic data. This program makes it possible to segregate essential data under different column headings, thus providing the means by which key information from a citation may be

Documentalist MN	Date of Preparation 7/31/72
EMIC Accession No.	012070
Agent	Methotrexate
Test Object	NMR I mice bearing hypertr iploid ascites utmor
Assay	Mitotic Index; Chromosome Aberrations; Host-Mediated Assay
Treatment Conditions	Agent in physiological solution injected i.p. 3 days following i.p. injection of 0.05 ml ascites; samples of ascites liquld extracted daily and examined cytologically
Agent Concentrations Tested	25–200 mg/kg
Treatment Time	18–282 hr
Treatment Temperature	
Reported Biological Effect	Large decrease in mitotic index (0.1% at 18 hr); increases in abnormal metaphases (80%), anaphases (60%), gradual recovery with time; chromatid breaks, exchanges; giant chromosomes; tripolarity
Author's Remarks	Occurrence of many chromosome abnormalities even after mitotic recovery (66 hr) significant; single-dose effects useful in determining interval between successive doses

FIGURE 4. EMIC's tabular abstracting format.

Methotrexate

EMIC NO.	ORGANISM	ASSAY SYSTEM	TREATMENT CONDITIONS	CONCENTRATION	EFFECT. CONC.	TIME	TEMP.	BIOLOGICAL EFFECT	AUTHOR'S REMARKS
012070	NMR I mice bearing hyper-triploid ascites tumor	Mitotic Index; Chromosome Aberrations; Host-Mediated Assay	Agent in physiological solution injected i.p. 3 days following i.p. injection of 0.05 ml ascities; sample of ascites liquid extracted daily and examined cytologically	25–200 mg/kg		18–282 hr	—	Large decrease in mitotic index (0.1% at 18 hr), increases in abnormal metaphases (80%), anaphases (60%), and dicentric bridges (60%), gradual recovery with time; chromatid breaks, exchanges; giant chromosomes; tripolarity	Occurrence of many chromosome abnormalities even after mitotic recovery (66 hr) significant; single-dose effects useful in determining interval between successive doses

FIGURE 5. Tabular computer output.

summarized. An example of the type of information indexed using this method and the format used to process it for the computer are illustrated in Fig. 4. Data abstracted in this manner are then associated by the computer with their appropriate compound. These data are then printed in table form as shown in Fig. 5. Each entry can be queried individually or in combination with data contained under other headings.

To supplement its search methods, EMIC has access to a computer "Name Match" program which is located at the Oak Ridge National Laboratory's Computing Center. This program makes it possible to locate synonyms of chemical compounds so they may be used as search aids. This capability lends an important dimension to our ability to retrieve mutagenic data.

EMIC is also considering the application of several other techniques for use in data control and retrieval. For instance, in order to bring about a better understanding of chemical–biological interactions, implementation of a substructural search program is being studied. This would make it possible for the user to draw correlations between chemically active groups and mutagenic activity, a capability that could provide some prognostic value concerning the mutagenic activity of newly created chemicals. In conjunction with these data, a program capable of relating the metabolic fate of chemical compounds in different organisms is also under examination.

Presently, EMIC issues annual indexed bibliographies of the chemical mutagenesis literature. These surveys have marked the beginning of central organization of chemical mutagenesis information. Future surveys of this type will include more elaborate indexes for agent and organism identification. As the frequency of publication increases, the issuance of these collections will become more frequent. Through EMIC's present and experimental programs designed for the automated retrieval of mutagenic data, the literature reporting the testing of chemicals for mutagenicity will be rendered more useful and more accessible.

V. SOME SUGGESTIONS FOR IMPROVING LITERATURE CONTROL

Since publications in this literature field are being processed for storage and retrieval by computers, there are several adaptations authors could incorporate into their manuscripts which would make them more accessible for an intended user. For instance, most computer-generated indexes are prepared from titles by one of the following three methods: (1) keywords are selected and indexed from titles (KWIC Indexes), (2) lists of titles are subject-classified and accompanied by their bibliographical reference, or (3) entire tables of contents of journals are reproduced. If the names of chemicals, organisms, or test systems as well as biological-effect descriptor terms are absent from a title, much valuable information will be missed when these indexes are printed. Of course, in some situations, the inclusion of these items in titles is not possible, e.g., because of the number of chemicals tested, evaluation in more than one assay system, or observation of multiple biological effects. In such situations, the use of the most adequate identifiers should be implemented. There are many examples which could be cited to illustrate this point, but one can see the problem by looking over indexes prepared by one of the three previously described methods.

Journal publishing houses can assist in literature control by making it standard procedure to put the following information on the beginning page of each article they print: name of journal, volume, pages covered by article, and year of publication. Most journals comply to some extent with this idea but not all. Standardization of this practice would help lessen citation errors.

Another item quite frequently misused in the literature is abbreviations. This is especially true in the case of chemicals, as there are but few compounds which have abbreviations universally used. When abbreviations are used, they should be identified, as most journals require, on the first page of the article—usually as a footnote.

Other illustrations could be presented as examples of points which would make the tasks of literature control easier, but they will not be taken up any further here. However, consideration and adaptation of those points that have been covered will make the job of automated and manual manipulation of the literature less difficult.

VI. CONCLUSION

The growth of science has created many specialized and complex literature fields. The successful exploitation of this literature is a measure of its organization and the amount of knowledge intended users have about it. This chapter has presented a review of the chemical mutagenesis literature area with the intention of making it more familiar to those who use it. Future trends in the publication of chemical mutageneisis information will be accented by continued growth due to testing priorities and technological advancement. Additionally, the publication of negative experimental results will become more common, thus adding a new dimension of information to this field. In order to keep pace with this expansion, new methods of documentation control and information dispersal will have to be implemented so that these data will be readily available to the scientific community.

VII. ACKNOWLEDGMENTS

I am indebted to Elizabeth Von Halle and Lynn Hawkins for their critical review of this manuscript and to Drs. H. V. Malling and D. G. Doherty for their helpful suggestions.

Author Index

Bold numbers refer to chapters in this volume.

Subject Index